DEDEKIND SUMS

By

HANS RADEMACHER

and

EMIL GROSSWALD

THE
CARUS MATHEMATICAL MONOGRAPHS

Published by

THE MATHEMATICAL ASSOCIATION OF AMERICA

THE CARUS MATHEMATICAL MONOGRAPHS are an expression of the desire of Mrs. Mary Hegeler Carus, and of her son, Dr. Edward H. Carus, to contribute to the dissemination of mathematical knowledge by making accessible at nominal cost a series of expository presentations of the best thoughts and keenest researches in pure and applied mathematics. The publication of the first four of these monographs was made possible by a notable gift to the Mathematical Association of America by Mrs. Carus as sole trustee of the Edward C. Hegeler Trust fund. The sales from these have resulted in the Carus Monograph Fund, and the Mathematical Association has used this as a revolving book fund to publish the succeeding monographs.

The exposition of mathematical subjects which the monographs contain are set forth in a manner comprehensible not only to teachers and students specializing in mathematics, but also to scientific workers in other fields, and especially to the wide circle of thoughtful people who, having a moderate acquaintance with elementary mathematics, wish to extend their knowledge without prolonged and critical study of the mathematical journals and treatises. The scope of this series includes also historical and biographical monographs.

The following monographs have been published:

No. 1. Calculus of Variations, by G. A. Bliss
No. 2. Analytic Functions of a Complex Variable, by D. R. Curtiss
No. 3. Mathematical Statistics, by H. L. Rietz
No. 4. Projective Geometry, by J. W. Young
No. 5. A History of Mathematics in America before 1900, by D. E. Smith and Jekuthiel Ginsburg (out of print)
No. 6. Fourier Series and Orthogonal Polynomials, by Dunham Jackson
No. 7. Vectors and Matrices, by C. C. MacDuffee
No. 8. Rings and Ideals, by N. H. McCoy
No. 9. The Theory of Algebraic Numbers, by Harry Pollard
No. 10. The Arithmetic Theory of Quadratic Forms, by B. W. Jones
No. 11. Irrational Numbers, by Ivan Niven
No. 12. Statistical Independence in Probability, Analysis and Number Theory, by Mark Kac
No. 13. A Primer of Real Functions, by Ralph P. Boas, Jr.
No. 14. Combinatorial Mathematics, by Herbert John Ryser
No. 15. Noncommutative Rings, by I. N. Herstein
No. 16. Dedekind Sums, by Hans Rademacher and Emil Grosswald

The Carus Mathematical Monographs

NUMBER SIXTEEN

DEDEKIND SUMS

By

HANS RADEMACHER,

Late Professor of Mathematics

University of Pennsylvania and Rockefeller University

and

EMIL GROSSWALD

Professor of Mathematics

Temple University

Published and Distributed by

THE MATHEMATICAL ASSOCIATION OF AMERICA

Library of Congress Catalog Number: 72-88698

INTRODUCTION

Professor Hans Rademacher was invited by the Mathematical Association of America to deliver the Earle Raymond Hedrick lectures at the 1963 summer meeting in Boulder, Colorado. Professor Rademacher chose the topic Dedekind Sums, and prepared a set of notes from which to deliver the lectures. However, a temporary illness prevented him from giving the lectures, and he prevailed upon his colleague and former student, Emil Grosswald, to make the presentation from the notes.

Professor Rademacher never edited for publication the notes he had prepared for the Hedrick lectures. However, after his death in 1969, the manuscript was found among his papers, with a signed request that Emil Grosswald edit and publish these lecture notes. Professor Grosswald responded affirmatively, and completed the editing of these somewhat fragmentary notes, which consisted of 45 handwritten pages and a sketch of a bibliography. In view of the extensive additions of proofs, historical remarks, and subsequent developments of the subject, made by Emil Grosswald, the Subcommittee on Carus Monographs of the Committee on Publications concluded that joint authorship of the finished monograph was appropriate. The Association is indebted to Professor Grosswald for his dedication in bringing this volume into print.

IVAN NIVEN, *Chairman of the Committee on Publications*

PREFACE

The Mathematical Association of America nominated Professor Hans Rademacher as Hedrick Lecturer for the summer meeting of 1963. As topics for these lectures, Professor Rademacher selected the "Dedekind Sums", a subject to which he had returned many times throughout his long and distinguished career and to which he had contributed immensely.

He prepared a set of notes (called in what follows the "Notes"), but a passing indisposition prevented him from delivering the lectures.

He recovered soon afterwards, was once more active in mathematics and wrote at least seven papers (among which there is also one on Dedekind sums) after 1963, but, for unknown reasons, failed to edit for publication his "Notes" of the Hedrick Lectures.

In September of 1967, Professor Rademacher was stricken by a cruel illness, from which he never recovered. After his death on February 7, 1969, the manuscript was found among his papers; on the first page had been added the following words, in Professor Rademacher's handwriting: "If I should be unable to publish these lectures, I wish to ask *Emil Grosswald* to edit and publish them. (signed) Hans Rademacher (dated) 8-th September, 1963."

Having had the privilege to be first Professor Rademacher's student, then, for many years, his colleague at the University of Pennsylvania and—I dare hope at least—his friend, there could be no question on my part about the acceptance of this assignment. This turned out to be more difficult than anticipated. The "Notes" consist of 45 handwritten pages and a sketch (2 pages) of a bibliography. The text is written in that specific, personal style, which defies imitation and makes anyone who had ever attended his lectures believe that, while reading, he actually hears the familiar voice of that great teacher.

Most proofs are suppressed—which is the reasonable thing for a Hedrick Lecture; instead, where a proof should appear, there is usually a reference to some paper containing it. The corresponding proof may have been used (either as actually published, or, more likely, as modified and streamlined by Professor Rademacher) in editing the "Notes" for publication. These references are often rather cryptic, such as [Rdm], or [Iseki], or even just [ ], [ ]. He *knew*, but we must guess which of the 3, 4, or more different proofs of Rademacher he had in mind, or what paper of which Iseki is meant (there are three active mathematicians of this name and two of them work on topics germane to the present one), or which of the several existing proofs by different mathematicians was to be used at a given place.

What was I supposed to do? It was impossible to take liberties with the text without risking to destroy what I consider one of its most valuable assets, its own characteristic style. On the other hand, it seemed indispensable to "flesh out" the "Notes" meant for oral presentation, by incorporating into them at least some proofs, some indications of the history of the subject matter and the impact it has had on subsequent developments.

After long hesitations and consultations with a referee and with Professor R. G. Bartle, Editor of the Carus Monographs, I decided on a compromise. I left the text of the "Notes" virtually unchanged, except for minor modifications required for clarity or for reasons of grammar*. Concerning the proofs that were missing and seemed desirable, in most cases I expanded existing, sketchy indications into proofs, if that could be done without overly long interruptions of the original text. In the other cases, I wrote up the proofs to the best of my ability, by making use to the largest possible extent of Professor Rademacher's own published papers. These proofs are collected into an Appendix and may be skipped at a first reading without impairing the clarity or continuity of the main text. I also wrote up a few pages on the history of the Dedekind sums and inserted these as a sixth Chapter.

It is in the nature of such a brief survey that not all valuable contributions to the theory of Dedekind sums could be mentioned. I have tried to select a representative sample, but wish to apologize to all those mathematicians who have contributed to this field and whose work is not mentioned here. I also wish to thank all those mathematicians who wrote to me and helped to make this historic survey as complete as possible.

Among the letters received there is one that requires

---

* In fact, I hesitated to do even that, because certain ways of expressing himself orally, while occasionally somewhat peculiar, were part of Professor Rademacher's charm and of his personal style of speaking. However, a careful study of his papers and books showed that in his published work he respected scrupulously the rules and customs of good style in written English. This convinced me that he himself would have made those minor modifications of his first draft, if he had edited the "Notes" for publication and, therefore, I proceeded accordingly.

special mention, namely that of Professor L. Carlitz. Professor Carlitz suggests that the generalizations of the Dedekind sums discussed in Chapter 5 should be renamed and be called henceforth Dedekind-Rademacher sums.

I consider this suggestion justified. I personally shall accept Professor Carlitz's suggestion and shall call these sums Dedekind-Rademacher sums in any paper I may write in the future. I also urge our colleagues to do the same. In the present book, however, these sums will still be called Dedekind sums. This is done not so much because the compound name is rather cumbersome, as because, in spite of all additions and modifications, the present book should still be considered as Professor Rademacher's own work, and this so very modest man would not have wished to call these sums by his own name.

In its present form, the book consists of five chapters based on the "Notes" of Professor Rademacher, a historic review and four notes assembled into an appendix. The historic remarks and the appendix are my own addition, but I must accept the responsibility for the whole book. However, it is my fervent hope that in spite of the many changes and additions, a sufficiently large portion of the main text has remained close to the original draft, so that the specific, unmistakable flavor, characteristic of all of Professor Rademacher's writings will not have been entirely lost. Whether this is the case or not, only the readers will be able to tell.

It is my pleasant duty to mention much help and assistance received. Without the cooperation of Professors I. Niven and R. G. Bartle of the Mathematical Association of America, the work on this book could not even have started. I also acknowledge with gratitude the help of Dr. Jean-

Louis Nicolas, who helped streamline some proofs and that of Mrs. G. Ballard and Mrs. M. Braid, who typed the manuscript with infinite patience and great care.

Finally, last but not least, my thanks go to Mrs. Irma Rademacher, who made available the manuscript and was helpful in every respect.

May this collective labor of love bring joy to many readers, as would have been the wish of that great mathematician and teacher, who was Professor Hans Rademacher.

EMIL GROSSWALD
Temple University

CONTENTS

CHAPTER 1

INTRODUCTION

At first glance, the Dedekind sums seem to be a highly specialized subject. These sums denoted by $s(h,k)$ are defined as follows: Let h, k be integers, $(h,k) = 1$, $k \geqq 1$; then we set

$$(1) \qquad s(h,k) = \sum_{\mu=1}^{k}\left(\left(\frac{h\mu}{k}\right)\right)\left(\left(\frac{\mu}{k}\right)\right).$$

Here and in the following the symbol $((x))$ is defined by

$$(2) \qquad ((x)) = \begin{cases} x - [x] - 1/2 & \text{if } x \text{ is not an integer,} \\ 0 & \text{if } x \text{ is an integer,} \end{cases}$$

with $[x]$ the greatest integer not exceeding x. This is the well-known sawtooth function of period 1 (see Figure 1),

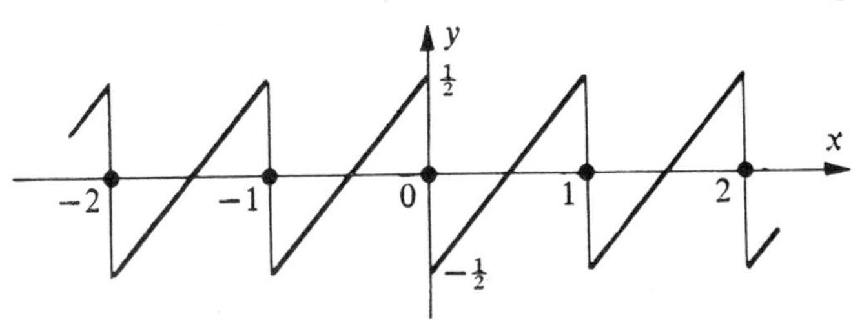

FIG. 1

which at the points of discontinuity takes the mean value between the limits from the right and from the left.

It is of course possible to define many sorts of sums or other expressions. In order to be fruitful, however, such definitions should not be made up arbitrarily, but should appear in a larger context. One may think, for example, of the Gaussian sums

$$G(h,k) = \sum_{\lambda=1}^{k} e^{\frac{2\pi ih}{k}\lambda^2},$$

about which there exists a whole literature inaugurated by Gauss himself. They appear naturally in the theory of quadratic residues, in cyclotomy, and in the theory of theta-functions, to name a few. Not only does such a full context ensure the mathematical usefulness and significance of the subject, but it also implies a rich structure of the subject itself.

The same can be said of the Dedekind sums. They are named after the mathematician Richard Dedekind (1831–1916) to whom our science owes so many great achievements, in particular the well-known theory of the Dedekind cut and the theory of ideals.

The Dedekind sums appear in Dedekind's study of the function

$$(3) \qquad \eta(\tau) = e^{\frac{\pi i\tau}{12}} \prod_{m=1}^{\infty} (1 - e^{2\pi im\tau}),$$

where $\operatorname{Im}\tau > 0$, as is needed for the convergence of the infinite product. This function is basic for the whole theory of elliptic functions and theta-functions. Using Jacobi's theory of elliptic functions, Dedekind showed (see [**14**]; also Chapter 4) that

$$(4) \qquad s(k, h) + s(h, k) = -\frac{1}{4} + \frac{1}{12}\left(\frac{h}{k} + \frac{1}{hk} + \frac{k}{h}\right).$$

Formula (4) is the famous reciprocity formula for Dedekind sums. Although derived in this context as a property of the function $\eta(\tau)$, it is simply an arithmetical formula and should thus be considered from the purely arithmetical point of view. This we shall do in the sequel; in fact, we shall only be concerned with the arithmetical nature of the Dedekind sums.

There exists quite a number of direct proofs of (4) in the literature. Chapter 2 is devoted to some of them, and the later chapters to arithmetical applications of the Dedekind sums.

CHAPTER 2

SOME PROOFS OF THE RECIPROCITY FORMULA

THEOREM 1. (Reciprocity Theorem). *Let h and k be two coprime integers. Then*

$$s(h,k) + s(k,h) = -\frac{1}{4} + \frac{1}{12}\left(\frac{h}{k} + \frac{1}{hk} + \frac{k}{h}\right). \tag{4}$$

A. First Proof. Let us begin with a proof which is a variant of one given by Ulrich Dieter [**16**]. We need first the

LEMMA 1. $$\sum_{\lambda \bmod k} \left(\left(\frac{\lambda + x}{k}\right)\right) = ((x)).$$

Proof: We consider the difference

$$D(x) = \sum_{\lambda \bmod k} \left(\left(\frac{\lambda + x}{k}\right)\right) - ((x)).$$

This function is obviously periodic in x with period 1. We may thus restrict our discussion to the range $0 \leqq x < 1$. If we choose, in particular, the residue system $\lambda = 0, 1, \cdots, k-1$, we have

$$D(0) = \sum_{\lambda=1}^{k-1}\left(\frac{\lambda}{k} - \frac{1}{2}\right) = \frac{k-1}{2} - \frac{k-1}{2} = 0.$$

Similarly, for $0 < x < 1$,

$$D(x) = \sum_{\lambda=0}^{k-1} \left(\frac{\lambda}{k} + \frac{x}{k} - \frac{1}{2}\right) - \left(x - \frac{1}{2}\right)$$

$$= \frac{k-1}{2} + x - \frac{k}{2} - x + \frac{1}{2} = 0.$$

This shows that $D(x) = 0$ for all values of x, and completes the proof of the Lemma.

LEMMA 2.

$$s(1,k) = -\frac{1}{4} + \frac{1}{6k} + \frac{k}{12}. \tag{5}$$

Proof: Formula (5) is the direct result of the definition:

$$s(1,k) = \sum_{\mu \bmod k} \left(\left(\frac{\mu}{k}\right)\right)^2$$

$$= \sum_{\mu=1}^{k-1} \left(\frac{\mu}{k} - \frac{1}{2}\right)^2 = \frac{1}{k^2} \sum_{\mu=1}^{k-1} \mu^2 - \frac{1}{k} \sum_{\mu=1}^{k-1} \mu + \frac{1}{4}(k-1)$$

$$= \frac{1}{k^2} \frac{(k-1)k(2k-1)}{6} - \frac{1}{k} \frac{k(k-1)}{2} + \frac{k-1}{4}$$

$$= -\frac{1}{4} + \frac{1}{6k} + \frac{k}{12},$$

as claimed.

We proceed now to the proof of the Reciprocity Theorem. We start with

$$S = s(h,k) + s(k,h)$$

$$= \sum_{\mu=0}^{k-1} \left(\left(\frac{\mu}{k}\right)\right) \left(\left(\frac{h\mu}{k}\right)\right) + \sum_{\nu=0}^{h-1} \left(\left(\frac{\nu}{h}\right)\right) \left(\left(\frac{k\nu}{h}\right)\right).$$

In the first sum, let $x = h\mu/k$, so that $(x+\lambda)/h = \mu/k + \lambda/h$, and proceed similarly in the second sum. Applying Lemma 1 in each sum we then obtain

$$S = \sum_{\mu=0}^{k-1} \left(\left(\frac{\mu}{k}\right)\right) \sum_{\nu=0}^{h-1} \left(\left(\frac{\nu}{h} + \frac{\mu}{k}\right)\right) + \sum_{\nu=0}^{h-1} \left(\left(\frac{\nu}{h}\right)\right) \sum_{\mu=0}^{k-1} \left(\left(\frac{\mu}{k} + \frac{\nu}{h}\right)\right)$$

$$= \sum_{\mu=0}^{k-1} \sum_{\nu=0}^{h-1} \left(\left(\frac{\mu}{k} + \frac{\nu}{h}\right)\right) \left\{ \left(\left(\frac{\mu}{k}\right)\right) + \left(\left(\frac{\nu}{h}\right)\right) \right\}.$$

Due to definition (2), the summands with $\mu = 0$ and $\nu = 0$ have to be considered separately. We have

$$(6) \quad S = \sum_{\mu=1}^{k-1} \sum_{\nu=1}^{h-1} \left(\left(\frac{\mu}{k} + \frac{\nu}{h}\right)\right) \left\{\frac{\mu}{k} + \frac{\nu}{h} - 1\right\}$$

$$+ \sum_{\nu=1}^{h-1} \left(\left(\frac{\nu}{h}\right)\right) \left(\frac{\nu}{h} - \frac{1}{2}\right) + \sum_{\mu=1}^{k-1} \left(\left(\frac{\mu}{k}\right)\right) \left(\frac{\mu}{k} - \frac{1}{2}\right),$$

where we suppressed the term with $\mu = \nu = 0$ (which equals zero anyway) in the original sum. Now, by Lemma 1 with $x = 0$,

$$\sum_{\mu \bmod k} \left(\left(\frac{\mu}{k}\right)\right) = 0 \quad \text{and} \quad \sum_{\nu \bmod h} \left(\left(\frac{\nu}{h}\right)\right) = 0,$$

and we can now recombine the sums in (6) into

$$S = \sum_{\mu=0}^{k-1} \sum_{\nu=0}^{h-1} \left(\left(\frac{\mu}{k} + \frac{\nu}{h}\right)\right) \left(\frac{\mu}{k} + \frac{\nu}{h} - 1\right).$$

We consider now the sum

$$(7) \qquad T = \sum_{\mu=0}^{k-1} \sum_{\nu=0}^{h-1} \left\{ \left(\frac{\mu}{k} + \frac{\nu}{h} - 1\right) - \left(\left(\frac{\mu}{k} + \frac{\nu}{h}\right)\right) \right\}^2$$

$$= \sum_{\mu=0}^{k-1}\sum_{\nu=0}^{h-1}\left(\frac{\mu}{k}+\frac{\nu}{h}-1\right)^2 - 2\sum_{\mu=0}^{k-1}\sum_{\nu=0}^{h-1}\left(\frac{\mu}{k}+\frac{\nu}{h}-1\right)\left(\left(\frac{\mu}{k}+\frac{\nu}{h}\right)\right)$$

$$+ \sum_{\mu=0}^{k-1}\sum_{\nu=0}^{h-1}\left(\left(\frac{\mu}{k}+\frac{\nu}{h}\right)\right)^2 = S_1 - 2S_2 + S_3,$$

say, where we recognize immediately that

$$S = S_2. \tag{8}$$

Moreover, S_1 is an elementary sum, namely

$$S_1 = \frac{h}{k^2}\sum_{\mu=0}^{k-1}\mu^2 + \frac{k}{h^2}\sum_{\nu=0}^{h-1}\nu^2 + hk + 2\frac{1}{hk}\sum_{\mu=0}^{k-1}\mu\sum_{\nu=0}^{h-1}\nu$$

$$- \frac{2h}{k}\sum_{\mu=0}^{k-1}\mu - \frac{2k}{h}\sum_{\nu=0}^{h-1}\nu,$$

which, after some straightforward computation, yields

$$S_1 = \frac{hk}{6} + \frac{h}{6k} + \frac{k}{6h} + \frac{1}{2}. \tag{9}$$

Next we observe that h and k are coprime; hence, when μ runs through a full residue system modulo k and ν through a full residue system modulo h, then the numbers $\rho = h\mu + k\nu$ run through a full residue system modulo hk.

It follows, taking into account Lemma 2, that

$$S_3 = \sum_{\rho \bmod hk}\left(\left(\frac{\rho}{hk}\right)\right)^2 = s(1, hk) = -\frac{1}{4} + \frac{1}{6hk} + \frac{hk}{12}.$$

We remember now that

$$z - ((z)) = \begin{cases} [z] + \frac{1}{2} & \text{for } z \text{ not an integer,} \\ [z] & \text{for integral } z. \end{cases}$$

In the double sum for T in (8), $\mu/k + \nu/h$ is an integer only for $\mu = \nu = 0$. Keeping this exception in mind, we obtain,

$$T = \sum_{\mu=0}^{k-1} \sum_{\nu=0}^{h-1} \left\{\left[\frac{\mu}{k} + \frac{\nu}{h}\right] - \frac{1}{2}\right\}^2 + \frac{3}{4},$$

where the correction term $\frac{3}{4}$ comes from the exceptional summand $\mu = \nu = 0$. It now follows that

$$\begin{aligned} T &= \sum_{\mu=0}^{k-1} \sum_{\nu=0}^{h-1} \left\{\left[\frac{\mu}{k} + \frac{\nu}{h}\right]^2 - \left[\frac{\mu}{k} + \frac{\nu}{h}\right] + \frac{1}{4}\right\} + \frac{3}{4} \\ &= \sum_{\mu=0}^{k-1} \sum_{\nu=0}^{h-1} \left[\frac{\mu}{k} + \frac{\nu}{h}\right]\left(\left[\frac{\mu}{k} + \frac{\nu}{h}\right] - 1\right) + \frac{1}{4} \cdot hk + \frac{3}{4}. \end{aligned}$$

Here we observe that within the range of values of μ and ν, we have $0 \leqq \mu/k + \nu/h < 2$, so that $[\mu/k + \nu/h] = 0$ or 1; consequently, the double sum vanishes and

$$T = \tfrac{1}{4}hk + \tfrac{3}{4}. \tag{10}$$

Putting together (7), (8), (9) and (10), we obtain

$$S = \frac{1}{2}(S_1 - T + S_3) = -\frac{1}{4} + \frac{1}{12}\left(\frac{h}{k} + \frac{1}{hk} + \frac{k}{h}\right),$$

and the Reciprocity Theorem is proved.

B. Second Proof. The sum $s(h,k)$ may be transformed as follows:

$$\begin{aligned} \sum_{\mu \bmod k} \left(\left(\frac{\mu}{k}\right)\right)\left(\left(\frac{h\mu}{k}\right)\right) &= \sum_{\mu=1}^{k-1} \left(\left(\frac{\mu}{k}\right)\right)\left(\left(\frac{h\mu}{k}\right)\right) \\ &= \sum_{\mu=1}^{k-1} \left(\frac{\mu}{k} - \frac{1}{2}\right)\left(\left(\frac{h\mu}{k}\right)\right) = \sum_{\mu=1}^{k-1} \frac{\mu}{k}\left(\left(\frac{h\mu}{k}\right)\right). \end{aligned}$$

Here the last equality holds, because $(h,k) = 1$ and, by Lemma 1 with $x = 0$,

$$\sum_{\mu=1}^{k-1}\left(\left(\frac{h\mu}{k}\right)\right) = \sum_{\mu=1}^{k-1}\left(\left(\frac{\mu}{k}\right)\right) = 0.$$

Thus the Reciprocity Theorem may be written as

$$\frac{1}{k}\sum_{\mu=1}^{k-1}\mu\left(\frac{h\mu}{k}-\left[\frac{h\mu}{k}\right]-\frac{1}{2}\right)+\frac{1}{h}\sum_{\nu=1}^{h-1}\nu\left(\frac{k\nu}{h}-\left[\frac{k\nu}{h}\right]-\frac{1}{2}\right)$$

$$= -\frac{1}{4}+\frac{1}{12}\left(\frac{h}{k}+\frac{1}{kh}+\frac{k}{h}\right).$$

In the first member appear some elementary sums, namely

$$\frac{1}{k}\sum_{\mu=1}^{k-1}\left(\mu^2\frac{h}{k}-\frac{\mu}{2}\right)=\frac{h}{k^2}\sum_{\mu=1}^{k-1}\mu^2-\frac{1}{2k}\sum_{\mu=1}^{k-1}\mu$$

$$=\frac{h}{k^2}\,\frac{(k-1)k(2k-1)}{6}-\frac{1}{2k}\,\frac{(k-1)k}{2}$$

$$=\frac{2h(k-1)(2k-1)}{12k}-\frac{3k(k-1)}{12k}=\frac{k-1}{12k}\{2h(2k-1)-3k\},$$

and the corresponding ones with h and k interchanged. If we replace these sums by their values and make some trivial simplifications, the Reciprocity Theorem to be proved becomes (see [**40**])

$$\text{(11)}\quad 12h\sum_{\mu=1}^{k-1}\mu\left[\frac{h\mu}{k}\right]+12k\sum_{\nu=1}^{h-1}\nu\left[\frac{k\nu}{h}\right]$$

$$=(h-1)(k-1)(8hk-h-k-1).$$

Now sums of square brackets suggest the counting of lattice points. In the classical proof of the quadratic reciprocity law one counts such lattice points in the plane. The sums that occur in (11) may be obtained by counting lattice points in three dimensional space. For that, consider the orthogonal parallelepiped $ABCDEFGH$ with edges h, k and hk, respectively (see Fig. 2). Let it be divided into three pyramids with common apex A. In the interior of the parallelepiped there are $(h-1)(k-1)(hk-1)$ lattice points. In the cleaving plane ACG there are no lattice points, since $(h,k) = 1$. But the planes AFG and AGH do contain lattice points; these can be counted, by projecting them down onto the plane $ABCD$. Their number is $(h-1)(k-1)$. Indeed, any lattice point P in the plane AFG, say, must have integral coordinates μ and ν, i.e., it must project onto a lattice point of the triangle ACB. Since $FB/AB = h$ it also follows that the ordinate of P is $\rho = h\mu$, an integer. Consequently, to each lattice point in ACB corresponds one (and quite obviously only one) lattice point above it in the plane AFG. Similarly to each lattice point in ACD corresponds one and only one lattice point in the

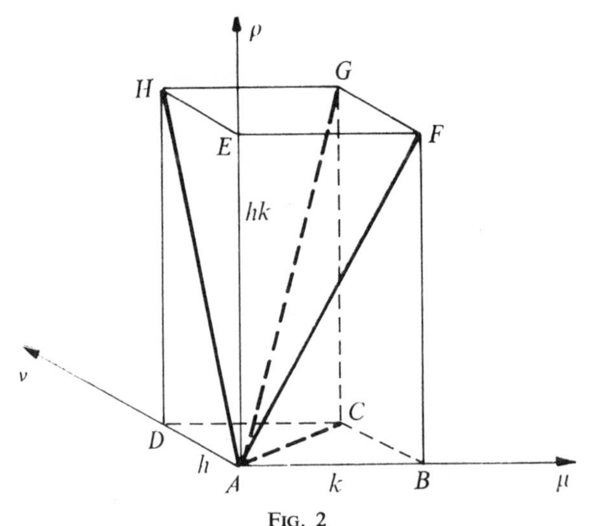

FIG. 2

plane AGH. No duplication of counting occurs, because there are no lattice points in ACG, and the result follows.

Now consider the two pyramids $A(BFGC)$ and $A(DCGH)$. We count their lattice points in planes parallel to their bases. In $A(BFGC)$ the rectangle in the plane at distance μ from A has base $h\mu/k$ and height μh and contains, for integral μ, $[h\mu/k]\mu h$ lattice points; hence, $A(BFGC)$ has $h\sum_{\mu=1}^{k-1}\mu[h\mu/k]$ lattice points, including those in the cleaving plane AFG. Similarly, there are $k\sum_{\nu=1}^{h-1}\nu[k\nu/h]$ lattice points in the pyramid $A(DCGH)$. In the pyramid $A(EFGH)$ we count the lattice points once more in layers parallel to its base $EFGH$ and obtain

$$\sum_{\rho=1}^{hk-1}\left[\frac{\rho}{h}\right]\left[\frac{\rho}{k}\right],$$

again counting those in the cleaving planes. If we add these three sums and observe that we have counted $(h-1)(k-1)$ lattice points twice (see the reasoning on page 10), we obtain

$$h\sum_{\mu=1}^{k-1}\mu\left[\frac{h\mu}{k}\right]+k\sum_{\nu=1}^{h-1}\nu\left[\frac{k\nu}{h}\right]+\sum_{\rho=1}^{hk-1}\left[\frac{\rho}{h}\right]\left[\frac{\rho}{h}\right]$$
$$-(h-1)(k-1)=(h-1)(k-1)(hk-1)$$

or

$$(12)\quad h\sum_{\mu=1}^{k-1}\mu\left[\frac{h\mu}{k}\right]+k\sum_{\nu=1}^{h-1}\nu\left[\frac{k\nu}{h}\right]+\sum_{\rho=1}^{hk-1}\left[\frac{\rho}{h}\right]\left[\frac{\rho}{k}\right]$$
$$=(h-1)(k-1)hk.$$

It turns out that the sum $\sum_{\rho=1}^{hk-1}[\rho/h][\rho/k]$ is of a simple nature, namely

$$(13)\quad \sum_{\rho=1}^{hk-1}\left[\frac{\rho}{h}\right]\left[\frac{\rho}{k}\right]=\frac{1}{12}(h-1)(k-1)(4hk+h+k+1)$$

holds. If we replace (13) in (12) a short computation yields (11), and the proof is complete.

It still remains to justify (13), which we now proceed to do. Let $(h,k)=1$ and denote the fractional part of x by $\{x\}=x-[x]$. Writing for simplicity Σ in place of $\Sigma_{\rho=1}^{hk-1}$, one obtains

$$U=\Sigma\left[\frac{\rho}{h}\right]\left[\frac{\rho}{k}\right]=\Sigma\left(\frac{\rho}{h}-\left\{\frac{\rho}{h}\right\}\right)\left(\frac{\rho}{k}-\left\{\frac{\rho}{k}\right\}\right)$$

$$=\frac{1}{hk}\Sigma\rho^2-\frac{1}{h}\Sigma\rho\left\{\frac{\rho}{k}\right\}-\frac{1}{k}\Sigma\rho\left\{\frac{\rho}{h}\right\}+\Sigma\left\{\frac{\rho}{h}\right\}\left\{\frac{\rho}{k}\right\}.$$

Here $\Sigma\rho^2=\frac{1}{6}(hk-1)hk(2hk-1)$. Next, if $\rho=rh+s$,

$$\Sigma\rho\left\{\frac{\rho}{h}\right\}=\sum_{s=0}^{h-1}\sum_{r=0}^{k-1}(rh+s)\frac{s}{h}=\sum_{s=0}^{h-1}s\sum_{r=0}^{k-1}r+\frac{1}{h}\sum_{s=0}^{h-1}\sum_{r=0}^{k-1}s^2$$

$$=\tfrac{1}{4}hk(h-1)(k-1)+\tfrac{1}{6}k(h-1)(2h-1),$$

so that

$$\begin{aligned}U&=\tfrac{1}{6}(hk-1)(2hk-1)-\tfrac{1}{4}h(h-1)(k-1)\\ &\quad-\tfrac{1}{6}(h-1)(2h-1)-\tfrac{1}{4}k(k-1)(h-1)-\tfrac{1}{6}(k-1)(2k-1)+V,\end{aligned}\tag{14}$$

where

$$V=\Sigma\left\{\frac{\rho}{h}\right\}\left\{\frac{\rho}{k}\right\}=\frac{1}{h}\sum_{s=0}^{h-1}s\sum_{r=0}^{k-1}\left\{\frac{rh+s}{k}\right\}.$$

Since $(h,k)=1$ it follows that if s is kept fixed and r runs through a complete set of residues $\bmod k$, also $rh+s$ runs through a complete set of residues modulo k. We thus obtain

$$\sum_{r=0}^{k-1}\left\{\frac{rh+s}{k}\right\}=\sum_{\nu=0}^{k-1}\frac{\nu}{k}=\frac{k-1}{2}$$

for every s, and

$$V = \frac{k-1}{2h} \sum_{s=0}^{h-1} s = \frac{(h-1)(k-1)}{4}.$$

If we now replace V in (14) by this value, we find

$$U = \frac{1}{12}(h-1)(k-1)(4hk+h+k+1),$$

and (13) is proved.

C. Third Proof. The previous proof* breaks up the symbol $((x))$ and so destroys an important periodicity. The following proofs will avoid this.

The arithmetical function $((\mu/k))$ has period k and can thus be expressed by a finite Fourier series. Let ζ be a primitive kth root of unity. Then

$$\left(\left(\frac{\mu}{k}\right)\right) = \sum_{m \bmod k} a_m \zeta^{\mu m} \tag{15}$$

will hold, with certain Fourier coefficients a_m. From (15) it follows that

$$\sum_{\mu \bmod k} \left(\left(\frac{\mu}{k}\right)\right) \zeta^{-\mu n} = \sum_{m \bmod k} a_m \sum_{\mu \bmod k} \zeta^{\mu(m-n)} = k a_n .$$

In particular, for $n = 0$, by Lemma 1,

$$a_0 = \frac{1}{k} \sum_{\mu \bmod k} \left(\left(\frac{\mu}{k}\right)\right) = 0. \tag{16a}$$

---

* The same remark holds also for another proof due to Mordell (see [36]), which also uses lattice points.

For $n \not\equiv 0 \pmod{k}$ we obtain

$$\text{(16b)} \quad a_n = \frac{1}{k} \sum_{\mu=1}^{k-1} \left(\frac{\mu}{k} - \frac{1}{2}\right) \zeta^{-\mu n} = \frac{1}{k^2} \sum_{\mu=1}^{k-1} \mu \zeta^{-\mu n} + \frac{1}{2k}.$$

Now $(1-x)(1+x+\cdots+x^{k-1}) = 1-x^k$, which yields, by differentiation

$$-(1+x+\cdots+x^{k-1}) + (1-x)(1+2x+\cdots+(k-1)x^{k-2}) = -kx^{k-1}.$$

If we multiply this equation by x and then replace x by ζ^{-n}, the first bracket vanishes and we have

$$(1-\zeta^{-n}) \sum_{\mu=1}^{k-1} \mu \zeta^{-\mu n} = -k.$$

This permits us to replace the sum in (16b) and, for $n \not\equiv 0 \pmod{k}$, we obtain

$$\text{(16c)} \qquad a_n = \frac{1}{k}\left(\frac{\zeta^n}{1-\zeta^n} + \frac{1}{2}\right).$$

Therefore, from (15), (16a) and (16c), we have now

$$\left(\left(\frac{\mu}{k}\right)\right) = \frac{1}{k} \sum_{n=1}^{k-1} \left(\frac{\zeta^n}{1-\zeta^n} + \frac{1}{2}\right) \zeta^{\mu n}.$$

From this we deduce that

$$s(h,k) = \frac{1}{k^2} \sum_{m=1}^{k-1} \left(\frac{\zeta^m}{1-\zeta^m} + \frac{1}{2}\right) \sum_{n=1}^{k-1} \left(\frac{\zeta^n}{1-\zeta^n} + \frac{1}{2}\right) \sum_{\mu \bmod k} \zeta^{(hm+n)\mu}.$$

The sum over μ vanishes, unless $n \equiv -hm \pmod{k}$, when it has the value k, so that

(17) $$s(h,k) = \frac{1}{k}\sum_{m=1}^{k-1}\left(\frac{\zeta^m}{1-\zeta^m}+\frac{1}{2}\right)\left(\frac{\zeta^{-hm}}{1-\zeta^{-hm}}+\frac{1}{2}\right).$$

This expression can now be used in several ways to obtain the reciprocity formula.

C$_1$. Firstly, a purely algebraic treatment is possible (see [**6**], [**43**], and [**47**]). In (17), ζ^m runs through all kth roots of unity with the exception of 1. Let us write ξ for the typical kth root of unity. Then (17) can be rewritten as

$$s(h,k) = \frac{1}{k}\sum_{\xi}{}'\left(\frac{\xi}{1-\xi}+\frac{1}{2}\right)\left(\frac{\xi^{-h}}{1-\xi^{-h}}+\frac{1}{2}\right),$$

where the prime $'$ indicates the omission of $\xi = 1$. By some simple manipulations, (see Note 1), this equation can be simplified to read

(18a) $$s(h,k) = -\frac{1}{k}\sum_{\xi}{}'\frac{1}{(\xi^h-1)(\xi-1)}+\frac{k-1}{4k}.$$

If η runs through all hth roots of unity except 1, we have in the same way

(18b) $$s(k,h) = -\frac{1}{h}\sum_{\eta}{}'\frac{1}{(\eta^k-1)(\eta-1)}+\frac{h-1}{4h}.$$

If we replace now in the Reciprocity Theorem $s(h,k)$ and $s(k,h)$ by their values (18a) and (18b), respectively, the identity to be proved reads

(19) $$\begin{aligned} &-\frac{1}{k}\sum_{\xi}{}'\frac{1}{(\xi^h-1)(\xi-1)}+\frac{k-1}{4k} \\ &-\frac{1}{h}\sum_{\eta}{}'\frac{1}{(\eta^k-1)(\eta-1)}+\frac{h-1}{4h} \\ &= -\frac{1}{4}+\frac{1}{12}\left(\frac{h}{k}+\frac{1}{hk}+\frac{k}{h}\right). \end{aligned}$$

The identity (19) may be considered as a relation between the k-th and the h-th roots of unity. This identity can be proved by elementary methods. The proof which follows [**6**] is based on the following two auxiliary identities:

$$(20)\quad \frac{1}{k}\sum_{\xi}{}' \frac{\xi}{(\xi^h-1)(\xi-1)}+\frac{1}{h}\sum_{\eta}{}' \frac{\eta}{(\eta^k-1)(\eta-1)}$$

$$=\frac{3(h+k-hk)-(h^2+k^2+1)}{12\,hk},$$

and

$$(21)\quad \frac{1}{k}\sum_{\xi}{}' \frac{1}{\xi^h-1}+\frac{1}{h}\sum_{\eta}{}' \frac{1}{\eta^k-1}=\frac{1}{2}\left(\frac{1}{h}+\frac{1}{k}\right)-1.$$

If we subtract (21) from (20), we obtain

$$(22)\quad \frac{1}{k}\sum_{\xi}{}' \frac{1}{(\xi^h-1)(\xi-1)}+\frac{1}{h}\sum_{\eta}{}' \frac{1}{(\eta^k-1)(\eta-1)}$$

$$=\frac{3}{4}-\frac{1}{12}\left(\frac{h+3}{k}+\frac{1}{hk}+\frac{k+3}{h}\right),$$

which is equivalent to (19). The proof of the Reciprocity Theorem has been reduced to that of (20) and (21), which we now proceed to indicate.

The first sum in (21) is

$$\sum_{\xi}{}' \frac{1}{\xi^h-1}=\sum_{\xi}{}' \frac{1}{\xi-1}$$

and is real, as can be seen by replacing ξ by $\bar{\xi}=\xi^{k-1}=\xi^{-1}$. With $\xi=e^{2\pi i/k}$ the sum becomes

$$\sum_{m=1}^{k-1} \frac{1}{e^{2\pi im/k}-1} = \sum_{m=1}^{k-1} \frac{e^{-\pi im/k}}{2i\sin(\pi m/k)}$$

$$= \frac{1}{2i}\sum_{m=1}^{k-1} \frac{\cos(\pi m/k) - i\sin(\pi m/k)}{\sin(\pi m/k)}$$

$$= \frac{1}{2i}\sum_{m=1}^{k-1} \cot\frac{\pi m}{k} - \frac{k-1}{2} = -\frac{k-1}{2},$$

because the sum is real. Consequently, the first member of (21) is

$$-\frac{k-1}{2k} - \frac{h-1}{2h} = -1 + \frac{1}{2}\left(\frac{1}{h} + \frac{1}{k}\right),$$

and this proves (21).

To prove (20), observe that $(h,k) = 1$ implies that the polynomials $p(x) = (x^k - 1)/(x-1)$ and $q(x) = (x^h - 1)/(x-1)$ are relatively prime; hence, there exist polynomials $f(x)$ and $g(x)$ of degrees at most $k-2$ and $h-2$, respectively, such that

$$q(x)f(x) + p(x)g(x) = 1. \tag{23}$$

For $x = \xi = e^{2\pi im/k}$ and $m \not\equiv 0 \pmod{k}$, $q(\xi)f(\xi) = 1$, or $f(\xi) = (\xi - 1)/(\xi^h - 1)$; and similarly $g(\eta) = (\eta - 1)/(\eta^k - 1)$. As $x \to 1$, (23) tends to

$$hf(1) + kg(1) = 1. \tag{24}$$

By Lagrange's interpolation formula we now obtain

$$f(x) = \frac{x^k - 1}{k}\left\{\frac{f(1)}{x-1} + \sum_{\xi}{}' \frac{\xi}{x-\xi}\,\frac{\xi - 1}{\xi^h - 1}\right\},$$

with a similar formula for $g(x)$. Substituting these in (23) and using also (24) we obtain

$$\frac{1}{k}\sum_{\xi}{}' \frac{\xi}{x-\xi}\,\frac{\xi-1}{\xi^h-1}+\frac{1}{h}\sum_{\eta}{}' \frac{\eta}{x-\eta}\,\frac{\eta-1}{\eta^k-1}$$

$$=\frac{x-1}{(x^k-1)(x^h-1)}-\frac{1}{hk(x-1)}.$$

We now set $x = 1 + t$ and expand both sides in powers of t. Equating the coefficients of t we obtain (20). This finishes the proof of the Reciprocity Theorem.

C$_2$. Returning to formula (17) for another approach, we write it as

$$s(h,k) = \frac{1}{4k}\sum_{m=1}^{k-1}\frac{1+\zeta^m}{1-\zeta^m}\,\frac{1+\zeta^{-hm}}{1-\zeta^{-hm}}. \tag{25}$$

Here ζ is again any primitive kth root of unity. If we specify now $\zeta = e^{2\pi i/k}$, then (25) becomes

$$s(h,k) = \frac{1}{4k}\sum_{m=1}^{k-1}\cot\frac{\pi m}{k}\cot\frac{\pi h m}{k}. \tag{26}$$

The Reciprocity Theorem (4) can now be derived in a manner analogous to that in the first proof (see [**16**], [**32**]) by means of

LEMMA 3. *If x is not an integer, then*

$$\sum_{m=1}^{k}\cot\frac{m+x}{k}\pi = k\cot x\pi. \tag{27}$$

Proof. We start from the decomposition into partial fractions

$$\frac{k}{z^k-1}=\sum_{l=1}^{k}\frac{\zeta^l}{z-\zeta^l},$$

with ζ a kth root of unity, as before. It follows that

$$\sum_{l=1}^{k} \frac{z+\zeta^l}{z-\zeta^l} = \sum_{l=1}^{k} \left(1 + 2\,\frac{\zeta^l}{z-\zeta^l}\right) = k + \frac{2k}{z^k-1} = k\,\frac{z^k+1}{z^k-1}.$$

If we set $z = e^{2\pi i\alpha}$ and replace l by $-m$, we obtain (27) with $x = \alpha k$.

We start now from (26) and have

$$S = s(h,k) + s(k,h)$$

$$= \frac{1}{4k}\sum_{m=1}^{k-1} \cot\frac{\pi m}{k}\cot\frac{\pi hm}{k} + \frac{1}{4h}\sum_{n=1}^{h-1} \cot\frac{\pi n}{h}\cot\frac{\pi kn}{h}.$$

Now we use (27) in order to obtain

$$S = \frac{1}{4k}\sum_{m=1}^{k-1} \cot\frac{\pi m}{k}\cdot\frac{1}{h}\sum_{n=1}^{h} \cot\pi\left(\frac{n}{h}+\frac{m}{k}\right)$$

$$+ \frac{1}{4h}\sum_{n=1}^{h-1} \cot\frac{\pi n}{h}\cdot\frac{1}{k}\sum_{m=1}^{k} \cot\pi\left(\frac{m}{k}+\frac{n}{h}\right)$$

$$= \frac{1}{4hk}\sum_{m=1}^{k-1}\sum_{n=1}^{h-1}\left(\cot\frac{\pi m}{k} + \cot\frac{\pi n}{h}\right)\cot\pi\left(\frac{m}{k}+\frac{n}{h}\right)$$

$$+ \frac{1}{4kh}\sum_{m=1}^{k-1}\cot\frac{\pi m}{k}\cot\pi\left(\frac{m}{k}+1\right) + \frac{1}{4hk}\sum_{n=1}^{h-1}\cot\frac{\pi n}{h}\cot\pi\left(1+\frac{n}{h}\right)$$

$$= \frac{1}{4kh}\left\{\sum_{m=1}^{k-1}\sum_{n=1}^{h-1}\left(\cot\frac{\pi m}{k} + \cot\frac{\pi n}{h}\right)\cot\pi\left(\frac{m}{k}+\frac{n}{h}\right)\right.$$

$$\left. + \sum_{m=1}^{k-1}\left(\cot\frac{\pi m}{k}\right)^2 + \sum_{n=1}^{h-1}\left(\cot\frac{\pi n}{h}\right)^2\right\}.$$

We recall that $\cot(\alpha+\beta)(\cot\alpha+\cot\beta) = \cot\alpha\cot\beta - 1$, so that the first sum becomes

$$\sum_{=1}^{k-1}\sum_{n=1}^{h-1} \cot\frac{\pi m}{k}\cot\frac{\pi n}{h} - (h-1)(k-1).$$

Using (26), we identify the last two sums as $4k\, s(1,k)$ and $4h\, s(1,h)$, respectively, so that

$$S = \frac{1}{4hk} \sum_{m=1}^{k-1} \sum_{n=1}^{h-1} \cot\frac{\pi m}{k} \cot\frac{\pi n}{h} - \frac{(h-1)(k-1)}{4hk}$$

$$+ \frac{1}{h} s(1,k) + \frac{1}{k} s(1,h).$$

The double sum vanishes, because $\sum_{m=1}^{k-1} \cot(\pi m/k) = 0$. Also, according to Lemma 2

$$s(1,k) = -\frac{1}{4} + \frac{k}{12} + \frac{1}{6k} \quad \text{and} \quad s(1,h) = -\frac{1}{4} + \frac{h}{12} + \frac{1}{6h},$$

so that

$$S = -\frac{1}{4} + \frac{1}{12}\left(\frac{h}{k} + \frac{1}{hk} + \frac{k}{h}\right).$$

This is again the Reciprocity Theorem.

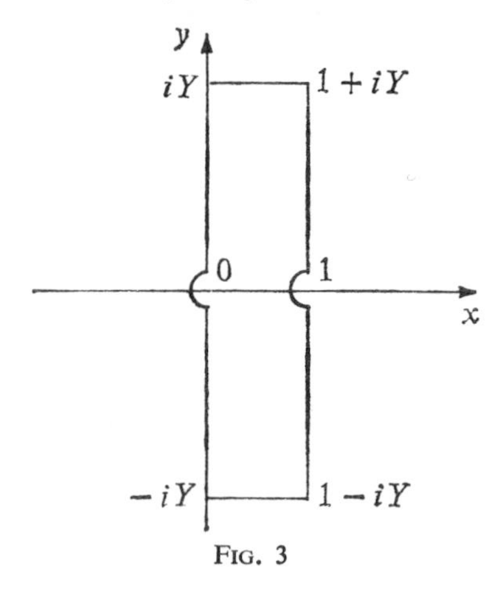

FIG. 3

C_3. Another way from (26) to the Reciprocity Theorem is by contour integration of the function

$$F(z) = \cot \pi z \cot \pi h z \cot \pi k z$$

around a certain contour $\mathscr{C}$. This contour (see Fig. 3) is obtained from the rectangle with vertices $1+iY, iY, -iY, 1-iY$, $Y > 0$, by making two small, semicircular, parallel indentations, which keep $z=0$ inside, and $z=1$ outside the contour. Here Y is a large, but otherwise arbitrary positive number. $F(z)$ has a triple pole at $z = 0$ and poles at $z = \lambda/k$ $(\lambda = 1,2,\cdots,k-1)$ and $z = \mu/h$ $(\mu = 1,2,\cdots,h-1)$. These are simple poles, because $(h,k) = 1$. Also, $F(z)$ is holomorphic on $\mathscr{C}$, so that, by Cauchy's theorem,

$$\frac{1}{2\pi i} \int_{\mathscr{C}} F(z) dz = S,$$

where S denotes the sum of the residues of $F(z)$ inside $\mathscr{C}$. Clearly, $F(z) = F(z+1)$; hence, the integrals along the vertical sides (inclusive of the indentations of $\mathscr{C}$) cancel, being taken in opposite directions. For $y \to \pm\infty$, $\cot \pi(x+iy) \to \mp i$; hence if $Y \to \infty$, $\int_{\mathscr{C}} F(z)dz \to -2i$. In fact, the integral is independent of Y, so that, for every $Y > 0$,

$$S = \frac{1}{2\pi i} \int_{\mathscr{C}} F(z)\, dz = -\frac{1}{\pi}.$$

We now pass to the computation of the residues. It is immediate that at $z = \lambda/h$ the residue is $(1/\pi h)\cot(\pi\lambda/h)$ $\cot(\pi k \lambda/h)$, and at $z = \mu/k$ it is $(1/\pi k)\cot(\pi\mu/k)\cot(\pi\mu h/k)$. Next, using the expansion $\cot z = 1/z - z/3 - \cdots$, it follows that in a neighborhood of $z = 0$,

$$F(z) = \frac{1}{\pi^3 hkz^3}\left(1 - \frac{\pi^2 z^2}{3} - \cdots\right)\left(1 - \frac{\pi^2 h^2 z^2}{3} - \cdots\right)$$
$$\times \left(1 - \frac{\pi^2 k^2 z^2}{3} - \cdots\right),$$

so that the residue of $F(z)$ at $z = 0$ is $(-1/3\pi)(h/k + 1/hk + k/h)$. Consequently,

$$S = -\frac{1}{3\pi}\left(\frac{h}{k} + \frac{1}{hk} + \frac{k}{h}\right) + \frac{1}{\pi h}\sum_{\lambda=1}^{h-1} \cot\frac{\pi\lambda}{h}\cot\frac{\pi\lambda k}{h}$$
$$+ \frac{1}{\pi k}\sum_{\mu=1}^{k-1} \cot\frac{\pi\mu}{k}\cot\frac{\pi\mu h}{k},$$

or, on account of (26),

$$S = \frac{4}{\pi}\left\{-\frac{1}{12}\left(\frac{h}{k} + \frac{1}{hk} + \frac{k}{h}\right) + s(h,k) + s(k,h)\right\}.$$

If we replace here S by its value $-1/\pi$ we obtain again the Reciprocity Theorem.

D. Fourth Proof. A most peculiar proof, with which we conclude this chapter, is based on a lemma about Stieltjes integrals. One can prove (see [**45**] or Note 2) the following

LEMMA 4. *Let $f(x)$, $g(x)$, and $q(x)$ be real valued functions of bounded variation in $a \leqq x \leqq b$, no two of which have any discontinuities in common. Then*

$$(28) \quad \int_a^b f(x)d(g(x)q(x))$$
$$= \int_a^b f(x)g(x)dq(x) + \int_a^b f(x)q(x)dg(x).$$

REMARK. For $f(x) = 1$, this yields simply the formula of integration by parts

$$(29)\qquad g(x)q(x)\Big|_a^b = \int_a^b g(x)dq(x) + \int_a^b q(x)dg(x).$$

Also the following lemma is known (see [**28**], pp. 170–171):

LEMMA 5. *For* $(h,k) = 1$,

$$(30)\qquad \int_0^1 ((hx))\,((kx))\,dx = 1/12hk.$$

We shall assume for a moment the validity of Lemma 5 and consider first the following application of (28):

$$(31)\quad I_\varepsilon = \int_\varepsilon^{1-\varepsilon} ((x))d(((hx))((kx)))$$
$$= \int_\varepsilon^{1-\varepsilon} ((x))((hx))d((kx)) + \int_\varepsilon^{1-\varepsilon} ((x))((kx))d((hx)),$$

with small $\varepsilon > 0$. The points 0 and 1 had to be omitted from the interval of integration, because there the three functions have a discontinuity in common.

In order to compute the right-hand side of (31) we have to take into account the discontinuities of the expressions under the integral signs. We observe that each jump is equal to -1 and obtain

$$I_\varepsilon = k\int_\varepsilon^{1-\varepsilon} ((x))((hx))dx - \sum_{\mu=1}^{k-1}\left(\left(\frac{\mu}{k}\right)\right)\left(\left(\frac{h\mu}{k}\right)\right)$$
$$+ h\int_\varepsilon^{1-\varepsilon} ((x))((kx))dx - \sum_{\nu=1}^{h-1}\left(\left(\frac{\nu}{h}\right)\right)\left(\left(\frac{\nu k}{h}\right)\right).$$

An application of (30) then yields

$$\lim_{\varepsilon\to 0} I_\varepsilon = \frac{k}{12h} + \frac{h}{12k} - s(h,k) - s(k,h). \tag{32}$$

On the other hand, one can transform I_ε by means of (29) (with $g(x) = ((x))$ and $q(x) = ((hx))((kx)))$ and then one obtains, using also Lemma 5, that

$$\lim_{\varepsilon\to 0} I_\varepsilon = \lim_{\varepsilon\to 0} (x - \tfrac{1}{2})((hx))((kx)) \Big|_{\varepsilon}^{1-\varepsilon} - \lim_{\varepsilon\to 0} \int_{\varepsilon}^{1-\varepsilon} ((hx))((kx))\, dx$$

$$= \left(\frac{1}{2}\right)^3 - \left(-\frac{1}{2}\right)^3 - \int_0^1 ((hx))((kx))dx = \frac{1}{4} - \frac{1}{12hk}.$$

This, together with (32), establishes again the Reciprocity Theorem.

We still have to justify the use of (30).

Proof of Lemma 5. Let $(h,k) = 1$; then

$$I = \int_0^1 ((hx))((kx))dx = \sum_{n=0}^{h-1} \int_{n/h}^{(n+1)/h} ((hx))((kx))dx.$$

With $x = y/h + n/h$, this becomes

$$I = \frac{1}{h} \sum_{n=0}^{h-1} \int_0^1 ((y+n)) \left(\left(\frac{ky}{h} + \frac{kn}{h}\right)\right) dy$$

$$= \frac{1}{h} \sum_{n=0}^{h-1} \int_0^1 ((y)) \left(\left(\frac{ky}{h} + \frac{kn}{h}\right)\right) dy$$

$$= \frac{1}{h} \int_0^1 ((y)) \sum_{n=0}^{h-1} \left(\left(\frac{ky}{h} + \frac{kn}{h}\right)\right) dy.$$

By Lemma 1 the sum equals $((ky))$, so that

$$I = \frac{1}{h}\int_0^1 ((y))((ky))dy.$$

Repeating the procedure on k, we obtain

$$\begin{aligned}\int_0^1 ((y))((ky))dy &= \frac{1}{k}\int_0^1 ((y))((y))dy \\ &= \frac{1}{k}\int_0^1 \left(y - \frac{1}{2}\right)^2 dy = \frac{1}{12k}\end{aligned}$$

and $I = 1/12hk$ as claimed.

REMARK. If $(h,k) = c > 1$, set $h = ca$, $k = cb$, with $(a,b) = 1$. Then

$$\begin{aligned}I &= \int_0^1 ((hx))((kx))dx = \int_0^1 ((acx))((bcx))dx \\ &= \frac{1}{c}\int_0^c ((ay))((by))dy,\end{aligned}$$

and, using the periodicity modulo 1 of the integrand, we obtain

$$I = \frac{1}{c}\cdot c\cdot\int_0^1 ((ay))((by))dy = \frac{1}{12ab} = \frac{c^2}{12hk}.$$

CHAPTER 3

ARITHMETIC PROPERTIES OF THE DEDEKIND SUMS

A. Elementary Properties. The reciprocity law of the Dedekind sums always contains two (and in some generalizations three and even more) Dedekind sums. We focus our attention now on a single Dedekind sum, its properties and its connections with other mathematical topics.

Since $((-x)) = -((x))$ it is clear that

$$s(-h,k) = -s(h,k) \tag{33a}$$

and

$$s(h,-k) = s(h,k). \tag{33b}$$

If we define h' by $hh' \equiv 1 \pmod{k}$, then

$$s(h',k) = s(h,k). \tag{33c}$$

Indeed,

$$s(h,k) = \sum_{\mu \bmod k} \left(\left(\frac{\mu}{k}\right)\right)\left(\left(\frac{h\mu}{k}\right)\right) = \sum_{\mu \bmod k} \left(\left(\frac{h'\mu}{k}\right)\right)\left(\left(\frac{hh'\mu}{k}\right)\right)$$

$$= \sum_{\mu \bmod k} \left(\left(\frac{h'\mu}{k}\right)\right)\left(\left(\frac{\mu}{k}\right)\right) = s(h',k).$$

Next, we may state the following:

THEOREM 2. *The denominator of* $s(h,k)$ *is a divisor of* $2k(3,k)$.

Proof of Theorem 2.

$$s(h,k) = \sum_{\mu=1}^{k-1} \left(\frac{\mu}{k} - \left[\frac{\mu}{k}\right] - \frac{1}{2}\right)\left(\frac{h\mu}{k} - \left[\frac{h\mu}{k}\right] - \frac{1}{2}\right)$$

$$= \frac{h}{k^2}\sum_{\mu=1}^{k-1}\mu^2 + \frac{A}{2k} + \frac{1}{4}\sum_{\mu=1}^{k-1} 1,$$

with A an integer, so that

$$s(h,k) = \frac{h(k-1)(2k-1)}{6k} + \frac{A}{2k} + \frac{k-1}{4}. \tag{34}$$

If k is even, $4 \mid 2k$; if k is odd, $k-1$ is even, and the last fraction has, after reduction, at most the denominator 2. In either case the denominator of the (reduced) last fraction divides $2k$, and the same holds for the second fraction. If $3 \nmid k$, then $(3,k) = 1$ and $2k(3,k) = 2k$. In this case $3 \mid (k-1)(2k-1)$, so that the factor 3 cancels in the first fraction, and the denominator is reduced to a factor of $2k$ (which might be $2k$ itself). If $3 \mid k$, then $2k(3,k) = 6k$, and in either case the denominator of the (reduced) first fraction divides $2k(3,k)$. The proof of the theorem is complete.

Clearly, $s(0,1) = s(1,1) = 0$, and zero is the only integer value which $s(h,k)$ can attain (see [**49**]). Indeed it follows from (4) that

$$12hk\,s(h,k) + 12hk\,s(k,h) = -3hk + h^2 + k^2 + 1. \tag{35}$$

We also know by Theorem 2 that if $\theta = (3,k)$, then the denominator of $s(h,k)$ divides $2\theta k$, so that (35) yields

$$(36) \qquad 12hk\,s(h,k) \equiv h^2 + 1 \pmod{\theta k}.$$

(a) Assume that $h^2 + 1 \equiv 0 \pmod k$; then the congruence $hh' \equiv 1 \pmod k$ has the solution $h' = -h$, and (see (33a) and (33c)) one obtains $s(h,k) = s(h',k) = s(-h,k) = -s(h,k)$, so that

$$s(h,k) = 0.$$

(b) Conversely, assume that $12s(h,k)$ is an integer. Then, if we recall that, by Theorem 2, $6h\,s(k,h)$ is always an integer, (35) shows that $h^2 + 1 \equiv 0 \pmod k$. This proves that one has actually the following stronger result: $s(h,k)$ is an integer (namely zero) if and only if $k \mid (h^2 + 1)$; if $k \nmid (h^2 + 1)$, then not even $12\,s(h,k)$ can be an integer.

The range of values of $s(h,k)$ is not fully known. Salié [**59**] proved that $s(h,k)$ always satisfies one of the following five congruences:

$$6ks(h,k) \equiv 0, \pm 1, \pm 3 \pmod 9.$$

If we write for a moment $s(h/k)$ instead of $s(h,k)$, the Dedekind sum becomes a function of a rational argument and has rational range. It can be shown (see [**49**]) that $s(h/k)$ is unbounded above and below in the neighborhood of every h/k. Whether the points $(h/k, s(h,k))$ lie everywhere dense in the plane is not yet known, nor does it seem to be known whether the range of $s(h,k)$ is dense on the real axis.

B. Farey Fractions. The set of all reduced rational fractions h/k, arranged in ascending order and with $1 \leqq k \leqq N$ and $0 \leqq h/k \leqq 1$, is called the Farey sequence of order N. We observe that the condition $(h,k) = 1$, excludes the value $h = 0$, except for $k = 1$.

Let $\phi(k)$ denote the Euler function, i.e., the number of positive integers not exceeding k and prime to k, or in symbols

$$\phi(k) = \sum_{\substack{n \leqq k \\ (n,k)=1}} 1 .$$

It is clear that for given k, the number of irreducible fractions h/k with $1 \leqq h \leqq k$ is $\phi(k)$, so that the total number of non-zero fractions in the Farey sequence of order N is

$$\Phi(N) = \sum_{k \leqq N} \phi(k).$$

It is well known (see, e.g. [**23**], p. 23) that if $h_1/k_1 < h_2/k_2$ are two consecutive fractions of a Farey sequence, then

$$\begin{vmatrix} h_1 & h_2 \\ k_1 & k_2 \end{vmatrix} = -1 .$$

This implies that

$$\begin{aligned} h_1 k_2 &\equiv -1 \pmod{k_1}, \\ h_2 k_1 &\equiv 1 \pmod{k_2}, \end{aligned}$$

and also that

$$\frac{h_2}{k_2} - \frac{h_1}{k_1} = \frac{1}{k_1 k_2}. \tag{37}$$

According to (33a) and (33c), we have thus

$$\begin{aligned} s(h_1, k_1) &= -s(k_2, k_1) \\ s(h_2, k_2) &= s(k_1, k_2). \end{aligned}$$

The reciprocity formula (4) then yields (see [**49**]):

$$(sh_2, k_2) - s(h_1, k_1) = -\frac{1}{4} + \frac{1}{12}\left(\frac{k_1}{k_2} + \frac{k_2}{k_1}\right) + \frac{1}{12k_1k_2}. \tag{38}$$

Summing (38) over all fractions of the sequence of order N and taking into account $s(0,1) = s(1,1) = 0$ and (37), we obtain

$$s(1,1) - s(0,1) = 0 = -\frac{1}{4}\Phi(N) + \frac{1}{12}\sum\left(\frac{k_1}{k_2} + \frac{k_2}{k_1}\right) + \frac{1}{12},$$

the sum Σ being extended over all pairs of consecutive Farey fractions h_1/k_1, h_2/k_2.

Now h/k belongs to a Farey sequence if and only if also h'/k, defined by $h'/k = 1 - h/k$, belongs to the same sequence, and then h/k and h'/k are symmetric with respect to the point $\frac{1}{2}$. It follows that

$$\sum\left(\frac{k_1}{k_2} + \frac{k_2}{k_1}\right) = 2\sum\frac{k_1}{k_2},$$

and we obtain the curious formula [**38**]

$$\sum_{k_1, k_2 \leqq N}\frac{k_1}{k_2} = \frac{3}{2}\Phi(N) - \frac{1}{2}.$$

If we remember also the asymptotic formula (see, e.g. [**23**, p. 268])

$$\Phi(N) = \frac{3}{\pi^2}N^2 + O(N\log N),$$

it follows, furthermore, that

$$\lim_{N\to\infty}\frac{1}{N^2}\sum_{k_1, k_2 \leqq N}\frac{k_2}{k_1} = \frac{9}{2\pi^2}.$$

C. Connection with the Jacobi Symbol. We recall the definition of the Legendre symbol (a/p): If p is a prime number and a an arbitrary integer, then

$$(a/p) = -1, 0, +1$$

accordingly as the congruence $x^2 \equiv a \pmod{p}$ has no solution, exactly one solution (this case happens if and only if $p \mid a$, with the unique solution $x \equiv 0 \pmod{p}$), or two distinct solutions modulo p. For $k = p_1 \cdots p_r$ (the primes $p_1, p_2, \cdots, p_r$ need not be distinct) an odd integer and $(h, k) = 1$, the Jacobi symbol (h/k) is defined by

$$\left(\frac{h}{k}\right) = \prod_{j=1}^{r} \left(\frac{h}{p_j}\right),$$

where (h/p_j) is the Legendre symbol.

The definition of the Dedekind sum, written explicitly, is

$$\begin{aligned} s(h,k) &= \sum_{\mu=1}^{k-1} \left(\frac{\mu}{k} - \frac{1}{2}\right) \left(\left(\frac{h\mu}{k}\right)\right) \\ &= \sum_{\mu=1}^{k-1} \frac{\mu}{k} \left(\frac{h\mu}{k} - \left[\frac{h\mu}{k}\right] - \frac{1}{2}\right), \end{aligned}$$

because, by Lemma 1, $\sum_{\mu=1}^{k-1}((h\mu/k)) = 0$. After some elementary computations we obtain

$$\begin{aligned} &(39) \quad 12ks(h,k) \\ &\quad = 2h(k-1)(2k-1) - 12 \sum_{\mu=1}^{k-1} \mu \left[\frac{h\mu}{k}\right] - 3k(k-1). \end{aligned}$$

Expressions like (39), usually divided by 2 or by 4, occur (in the theory of modular functions) as exponents of -1; therefore it is of interest to find the value of this expression modulo 8. We recall that k is odd, so that $k^2 \equiv 1 \pmod{8}$, and

$$\begin{aligned} 12ks(h,k) &\equiv 2h(2k^2 - 3k + 1) - 3k^2 + 3k - 4T \\ &\equiv (1-k)(6h-3) - 4T \pmod{8}. \end{aligned}$$

Here $T = \sum_{\mu=1}^{k-1} \mu[h\mu/k]$, and it remains to find $T \bmod 2$. For that, the even values of μ are irrelevant; hence,

$$(40) \quad T \equiv \sum_{\nu=1}^{(k-1)/2} (2\nu-1)\left[\frac{(2\nu-1)h}{k}\right] \equiv \sum_{\nu=1}^{(k-1)/2}\left[\frac{(2\nu-1)h}{k}\right]$$

$$\equiv \sum_{\mu=1}^{k-1}\left[\frac{\mu h}{k}\right] - \sum_{\mu=1}^{(k-1)/2}\left[\frac{2\mu h}{k}\right] \pmod 2 .$$

If $\{x\} = x - [x]$ denotes the fractional part of x, then

$$(41) \quad \begin{aligned} \sum_{\mu=1}^{k-1}\left[\frac{\mu h}{k}\right] &= \sum_{\mu=1}^{k-1}\left(\frac{\mu h}{k} - \left\{\frac{\mu h}{k}\right\}\right) \\ &= \frac{h}{k}\,\frac{k(k-1)}{2} - \sum_{\nu=1}^{k-1}\frac{\nu}{k} = \frac{(h-1)(k-1)}{2}. \end{aligned}$$

The next to the last equality follows from the fact that if $(h,k) = 1$ and μ runs through a complete set of residues $\bmod k$, then so does also μh. In order to estimate the last sum that occurs in (40) we need the following:

LEMMA 6. *If* $m = \sum_{\mu=1}^{(k-1)/2}[2\mu h/k]$ *and* (h/k) *denotes the Jacobi symbol, then* $(h/k) = (-1)^m$, *so that* $m \equiv \frac{1}{2}((h/k)-1)$ $\pmod 2$.

This Lemma is known (see [**44**], p. 397) but its proof is not easily accessible. For that reason we shall sketch it here.

Gauss' Lemma states (see e.g., [**22**], p. 71) that if k is an odd prime and $(k,h) = 1$, then the Legendre symbol (h/k) satisfies $(h/k) = (-1)^m$, where m is the number of least positive remainders exceeding $k/2$ in the sequence $\mu h \pmod k$, $\mu = 1,2,\cdots,(k-1)/2$ (equivalently, m is the number of absolutely least residues of $\mu h \pmod k$, that are negative). The generalization to arbitrary odd, positive k

with $(k, h) = 1$ offers no particular difficulties and may be found, e.g., in Bachmann's book [**3**], p. 144–148. According to Bachmann, this generalization is due to E. Schering [**60**]. To obtain the present result (and here we follow a footnote in [**44**, page 397]), observe that if $\mu h = \lambda k + r_\mu$, with $k/2 < r_\mu < k$, then $2\mu h = (2\lambda + 1)k + (2r_\mu - k)$, where $0 < 2r_\mu - k < k$. Consequently, $2\mu h/k = 2\lambda + 1 + \theta$, where $0 < \theta = (2r_\mu - k)/k < 1$ and $[2\mu h/k] = 2\lambda + 1$ is odd. Also the converse holds, because if $[2\mu h/k]$ is odd, this implies that $0 < \theta = (2r_\mu - k)/k < 1$, so that $k/2 < r_\mu < k$. The sum

$$\sum_{\mu=1}^{(k-1)/2} \left[\frac{2\mu h}{k}\right]$$

taken modulo 2 counts the number of odd $[2\mu h/k]$, i.e., the number of μ for which $k/2 < r_\mu < k$ holds, and that is precisely m. Consequently,

$$m \equiv \sum_{\mu=1}^{(k-1)/2} \left[\frac{2\mu h}{k}\right] \pmod 2,$$

as claimed.

If we now combine (40) with Lemma 6, we obtain $T \equiv \frac{1}{2}((h-1)(k-1) + (h/k) - 1) \pmod 2$. It follows that

$$\begin{aligned} 12k\,s(h,k) &\equiv (1-k)(6h-3) - 2(h-1)(k-1) - 2\left(\frac{h}{k}\right) + 2 \\ &\equiv (1-k)(8h-5) + 2 - 2\left(\frac{h}{k}\right) \equiv 5k - 3 - 2\left(\frac{h}{k}\right) \\ &\equiv k + 1 + 4(k-1) - 2\left(\frac{h}{k}\right) \pmod 8, \end{aligned}$$

so that

(42) $$12k\, s(h,k) \equiv k + 1 - 2\left(\frac{h}{k}\right) \pmod 8,$$

a formula due to Dedekind (see [**15**, § 6]).

It is now easy to obtain the Quadratic Reciprocity Theorem for the Jacobi symbol as a corollary of the Reciprocity Theorem (4) for Dedekind sums. Indeed, one obtains from (42)

$$12hk(s(h,k) + s(k,h)) \equiv 2hk + h + k - 2\left\{h\left(\frac{h}{k}\right) + k\left(\frac{k}{h}\right)\right\} \pmod 8.$$

On account of (4) this becomes, for odd h and k,

(43) $$5hk + h + k - 3 \equiv 2\left\{h\left(\frac{h}{k}\right) + k\left(\frac{k}{h}\right)\right\} \pmod 8.$$

Let us assume now that $k = 4m + 1$; then (43) becomes after some simplification $(2m-1)(h+1) \equiv h(h/k) + (k/h) \pmod 4$, or, since $2(h+1) \equiv 0 \pmod 4$,

$$h\left(1 + \left(\frac{h}{k}\right)\right) + \left(1 + \left(\frac{k}{h}\right)\right) \equiv 0 \pmod 4.$$

If $h \equiv 1 \pmod 4$, then $(k/h) + (h/k) \equiv \pm 2 \pmod 4$; if $h \equiv -1 \pmod 4$, then $(h/k) \equiv (k/h) \pmod 4$. But (h/k) and (k/h) can take only the values ± 1, so that in all these cases $(h/k) = (k/h)$. Exchanging h with k it follows that if either $h \equiv 1 \pmod 4$ or $k \equiv 1 \pmod 4$, then $(h/k) = (k/h)$. On the other hand, if $h \equiv k \equiv 3 \pmod 4$, then (43) becomes

$$\left(\frac{h}{k}\right) + \left(\frac{k}{h}\right) \equiv 0 \pmod 4,$$

so that $(h/k) = -(k/h)$. This finishes the proof of

THEOREM 3. (Quadratic Reciprocity Theorem for the Jacobi Symbol). *For odd, coprime integers* h *and* k,

$$\left(\frac{h}{k}\right)\left(\frac{k}{h}\right) = (-1)^{\frac{h-1}{2}\frac{k-1}{2}}$$

D. Zolotareff's Theorem. Consider the row

$$h, 2h, 3h, \cdots, (k-1)h, \tag{44}$$

and replace each entry by its smallest positive residue modulo k. The sequence becomes

$$r_1, r_2, \cdots, r_{k-1}, \tag{45}$$

and since $(h,k) = 1$, (45) is simply a permutation of $1, 2, 3, \cdots, k-1$.

We want to study the inversions in (45), i.e., the number of times a larger entry precedes a smaller one; indeed following E. Zolotareff, this number of inversions is related to the Legendre-Jacobi symbol. Let us denote that number of inversions in (45) by $I(h,k)$. Following C. Meyer [**32**] we shall be able to express $I(h,k)$ with the help of the Dedekind sum $s(h,k)$.

If we define $I_\mu(h,k)$ as the number of elements in $r_1, r_2, \cdots, r_{\mu-1}$ which exceed r_μ, then clearly

$$I(h,k) = \sum_{\mu=1}^{k-1} I_\mu(h,k) \tag{46}$$

with $I_1(h,k) = 0$. Now

$$r_\lambda = h\lambda - \left[\frac{h\lambda}{k}\right]k, \quad r_\mu = h\mu - \left[\frac{h\mu}{k}\right]k,$$

so that $r_\lambda > r_\mu$ means $h\lambda - [h\lambda/k]k > h\mu - [h\mu/k]k$, i.e.,

$[h\mu/k]-[h\lambda/k]>h(\mu-\lambda)/k>[h(\mu-\lambda)/k]$. Similarly, if $r_\lambda < r_\mu$, then

$$\left[\frac{h\mu}{k}\right]-\left[\frac{h\lambda}{k}\right]<\frac{h(\mu-\lambda)}{k}<\left[\frac{h(\mu-\lambda)}{k}\right]+1.$$

Hence, if $r_\lambda > r_\mu$, then

$$\left[\frac{h\mu}{k}\right]-\left[\frac{h\lambda}{k}\right]-\left[\frac{h(\mu-\lambda)}{k}\right]>0;$$

and, if $r_\lambda < r_\mu$, then

$$\left[\frac{h\mu}{k}\right]-\left[\frac{h\lambda}{k}\right]-\left[\frac{h(\mu-\lambda)}{k}\right]<1.$$

Now $[\alpha+\beta]-[\alpha]-[\beta]$ can only be 1 or 0 (see [**22**, p. 85]), so that $r_\lambda > r_\mu$ implies

$$\left[\frac{h\mu}{k}\right]-\left[\frac{h\lambda}{k}\right]-\left[\frac{h(\mu-\lambda)}{k}\right]=1,$$

and $r_\lambda < r_\mu$ implies

$$\left[\frac{h\mu}{k}\right]-\left[\frac{h\lambda}{k}\right]-\left[\frac{h(\mu-\lambda)}{k}\right]=0.$$

It follows that

$$\sum_{\lambda=1}^{\mu-1}\left(\left[\frac{h\mu}{k}\right]-\left[\frac{h\lambda}{k}\right]-\left[\frac{h(\mu-\lambda)}{k}\right]\right)$$

counts exactly the number of terms in $\{r_1, r_2, \cdots, r_{\mu-1}\}$ that exceed r_μ; i.e., the sum is equal to $I_\mu(h,k)$ by the very definition of $I_\mu(h,k)$. By virtue of (46) we obtain

$$I(h,k)=\sum_{\mu=1}^{k-1}\sum_{\lambda=1}^{\mu-1}\left(\left[\frac{h\mu}{k}\right]-\left[\frac{h\lambda}{k}\right]-\left[\frac{h(\mu-\lambda)}{k}\right]\right).$$

For purposes of simplification we add the vanishing terms corresponding to $\lambda = \mu$ and obtain

$$I(h,k) = \sum_{\mu=1}^{k-1} \sum_{\lambda=1}^{\mu} \left(\left[\frac{h\mu}{k}\right] - \left[\frac{h\lambda}{k}\right] - \left[\frac{h(\mu-\lambda)}{k}\right] \right)$$

$$= \sum_{\mu=1}^{k-1} \mu \left[\frac{h\mu}{k}\right] - \sum_{\mu=1}^{k-1} \sum_{\lambda=1}^{\mu} \left[\frac{h\lambda}{k}\right] - \sum_{\mu=1}^{k-1} \sum_{\rho=0}^{\mu-1} \left[\frac{h\rho}{k}\right]$$

$$= \sum_{\mu=1}^{k-1} \mu \left[\frac{h\mu}{k}\right] - 2 \sum_{1 \leqq \lambda \leqq} \sum_{\mu \leqq k-1} \left[\frac{h\lambda}{k}\right] + \sum_{\mu=1}^{k-1} \left[\frac{h\mu}{k}\right]$$

$$= \sum_{\mu=1}^{k-1} \mu \left[\frac{h\mu}{k}\right] - 2 \sum_{\lambda=1}^{k-1} (k-\lambda) \left[\frac{h\lambda}{k}\right] + \sum_{\mu=1}^{k-1} \left[\frac{h\mu}{k}\right],$$

and thus

$$I(h,k) = 3 \sum_{\mu=1}^{k-1} \mu \left[\frac{h\mu}{k}\right] - (2k-1) \sum_{\mu=1}^{k-1} \left[\frac{h\mu}{k}\right]. \tag{47}$$

We already found (see (41)) that for $(h,k) = 1$

$$\sum_{\mu=1}^{k-1} \left[\frac{h\mu}{k}\right] = \frac{(h-1)(k-1)}{2},$$

and also that the first sum in (47) can be expressed with the help of $s(h,k)$ by (39).

Making the corresponding substitutions, we obtain

$$I(h,k) = -3k\,s(h,k) + \tfrac{1}{4}(k-1)(k-2), \tag{48}$$

which is a formula of independent interest.

Now let k be odd. Then, on account of (42), (48) shows that

$$I(h,k) \equiv -\frac{k+1}{4} + \frac{1}{2}\left(\frac{h}{k}\right) + \frac{1}{4}(k-1)(k-2) \pmod 2$$

or

$$I(h,k) \equiv \frac{1}{2}\left\{\left(\frac{h}{k}\right) - 1\right\} + \frac{1}{4}(k^2-1) - k+1 \pmod 2$$

$$\equiv \frac{1}{2}\left\{\left(\frac{h}{k}\right) - 1\right\} \pmod 2.$$

Thus we obtain Zolotareff's Theorem:

THEOREM 4. (See [**64**] and [**58**]): *For odd k and* $(h,k)=1$,

$$(-1)^{I(h,k)} = \left(\frac{h}{k}\right). \tag{49}$$

In the particular cases $h = -1$ and $h = 2$, the number $I(h,k)$ of inversions is immediately computed from the corresponding sequences (45). For $h = -1$ we have $k-1, k-2, \cdots, 1$, so that

$$I(-1,k) = 1 + 2 + \cdots + (k-2) = \frac{(k-2)(k-1)}{2}.$$

For $h = 2$, (45) becomes $2, 4, \cdots, k-1, 1, 3, 5, \cdots, k-2$, and

$$I(2,k) = \frac{k-1}{2} + \left(\frac{k-1}{2} - 1\right) + \left(\frac{k-1}{2} - 2\right) + \cdots + 1$$

$$= \frac{k^2-1}{8}.$$

The already mentioned particular cases $h = -1$ and $h = 2$ of Zolotareff's Theorem (49) yield the well-known "supplementary" theorems of quadratic reciprocity:

$$\left(\frac{-1}{k}\right) = (-1)^{(k-1)/2}, \text{ and } \left(\frac{2}{k}\right) = (-1)^{(k^2-1)/8}.$$

Finally, if also h is positive and odd, we obtain from (48) and (4),

$$\begin{aligned} hI(h,k) + kI(k,h) &= -3hk(s(h,k) + s(k,h)) \\ &\quad + \tfrac{1}{4}(h(k-1)(k-2) + k(h-1)(h-2)) \\ &= -3hk\left\{-\frac{1}{4} + \frac{1}{12}\left(\frac{h}{k} + \frac{1}{hk} + \frac{k}{h}\right)\right\} \\ &\quad + \tfrac{1}{4}(hk^2 + h^2k - 6hk + 2h + 2k) \\ &= \tfrac{1}{4}(hk^2 + h^2k - 3hk - h^2 - k^2 + 2h \\ &\qquad + 2k - 1), \end{aligned}$$

or

$$hI(h,k) + kI(k,h) = \tfrac{1}{4}(h-1)(k-1)(h+k-1), \tag{50}$$

a formula of Salié ([**32**], p. 163).

This, of course, together with (49) once more implies immediately the quadratic reciprocity law of the Jacobi symbol

$$\left(\frac{h}{k}\right)\left(\frac{k}{h}\right) = (-1)^{\frac{h-1}{2}\frac{k-1}{2}}.$$

E. Lattice Points. We have seen that the discussion of the Dedekind sums is closely connected with the enumeration of lattice points in certain triangles, pyramids, and parallelepipeds.

The number of lattice points in a tetrahedron has been related to Dedekind sums by L. J. Mordell [**37**]. He proved the following result:

THEOREM 5. *Let a, b, and c be pairwise coprime, positive integers and let $N_3(a,b,c)$ be the number of lattice points in the tetrahedron*

$$(51) \quad 0 \leqq x < a, 0 \leqq y < b, 0 \leqq z < c, \; 0 < \frac{x}{a} + \frac{y}{b} + \frac{z}{c} < 1;$$

then

$$N_3(a,b,c) = -(s(bc,a) + s(ca,b) + s(ab,c)) + \tfrac{1}{6}abc$$
$$+ \frac{1}{4}(bc + ca + ab) + \frac{1}{4}(a + b + c) + \frac{1}{12}\left(\frac{bc}{a} + \frac{ca}{b} + \frac{ab}{c}\right)$$
$$+ \frac{1}{12abc} - 2.$$

Mordell's investigation starts with the elegant formula (here and in much of what follows we write simply N_3 for $N_3(a,b,c)$)

$$(52) \quad 2N_3 = \sum_{x,y,z}{}' \left(\left[\frac{x}{a} + \frac{y}{b} + \frac{z}{c}\right] - 1\right)\left(\left[\frac{x}{a} + \frac{y}{b} + \frac{z}{c}\right] - 2\right),$$

where the prime $'$ indicates the omission of the term $x = y = z = 0$, in accordance with the conditions (51) of the theorem. Indeed, the points with $0 < (x/a) + (y/b) + (z/c) < 1$ contribute each a summand 2, whereas those with

$$1 < \frac{x}{a} + \frac{y}{b} + \frac{z}{c} < 3$$

contribute 0. It may be observed here that

$$\frac{x}{a} + \frac{y}{b} + \frac{z}{c} = 1 \text{ or } 2$$

is impossible, since these equalities would imply

$$bcx + cay + abz \equiv 0 \pmod{abc},$$

and thus, simultaneously

$$x \equiv 0 \pmod{a}, \; y \equiv 0 \pmod{b}, \; z \equiv 0 \pmod{c},$$

which is excluded by the conditions of summation.

To get to the Dedekind sums, we set $(x/a) + (y/b) + (z/c) = E$ and write (52) as

$$2N_3 = \sum_{x,y,z}{}' \left\{E - ((E)) - \frac{3}{2}\right\} \left\{E - ((E)) - \frac{5}{2}\right\}$$

$$= \sum_{x,y,z}{}' (\{E - ((E)) - 2\}^2 - \tfrac{1}{4}) = \sum_{x,y,z}{}' \{E - ((E)) - 2\}^2 - \tfrac{1}{4}(\sum_{x,y,z}{}' 1)$$

$$= \sum_{x,y,z}{}' (E-2)^2 - 2 \sum_{x,y,z} (E-2)((E)) + \sum_{x,y,z} ((E))^2 - \tfrac{1}{4}(abc-1).$$

We have omitted the prime at the second and third sum, because for $x = y = z = 0$ also $E = 0$, so that $((E)) = 0$ and the corresponding terms vanish. We denote the last three sums by A, B, and C, respectively, so that (52) now reads

$$2N_3 = A - 2B + C - \tfrac{1}{4}(abc - 1). \tag{53}$$

We observe that if we want to suppress the prime of the summation sign Σ' in A, this amounts to the addition of the term with $x = y = z = 0$, i.e., to the addition of 4 to A so that we have

$$A + 4 = \sum_{x,y,z} \left(\frac{x}{a} + \frac{y}{b} + \frac{z}{c} - 2\right)^2.$$

The computation of the sum is elementary but lengthy (see Note 3) and yields

$$A = \frac{1}{2}\left(abc + bc + ca + ab + a + b + c + \frac{1}{3}\left(\frac{bc}{a} + \frac{ca}{b} + \frac{ab}{c}\right) - 8\right).$$

Further,

$$B = \sum_{x,y,z} (E-2)((E)) = \sum_{x,y,z} E((E)) - 2 \sum_{x,y,z} ((E)).$$

Here the last sum vanishes by Lemma 1, because $E = (bcx + cay + abz)/abc = t/abc$, say, and t runs through a full residue set modulo abc. when x, y, and z run through full residue sets modulo a, b, and c, respectively. Therefore,

$$B = \sum_{x,y,z} \frac{x}{a}((E)) + \sum_{x,y,z} \frac{y}{b}((E)) + \sum_{x,y,z} \frac{z}{c}((E)).$$

Here

$$\sum_{y,z} \frac{x}{a}((E)) = \frac{x}{a} \sum_{y,z} \left(\left(\frac{x}{a} + \frac{cy+bz}{bc}\right)\right) = \frac{x}{a} \sum_{\mu=1}^{bc} \left(\left(\frac{x}{a} + \frac{\mu}{bc}\right)\right).$$

By Lemma 1, the sum equals $((bc \cdot x/a))$, so that

$$\sum_{x,y,z} \frac{x}{a}((E)) = \sum_{0 \leqq x < a} \frac{x}{a} \left(\left(\frac{bc \cdot x}{a}\right)\right) = s(bc, a).$$

The second equality follows directly from (1) (see Chapter **2, B**). Handling the other two sums in the same way we obtain $B = s(bc,a) + s(ca,b) + s(ab,c)$. Finally,

$$C = \sum_{x,y,z} ((E))^2 = \sum_{t \bmod abc} \left(\left(\frac{t}{abc}\right)\right)^2$$

$$= s(1, abc) = -\frac{1}{4} + \frac{1}{6abc} + \frac{abc}{12},$$

by Lemma 2. Substituting these values of A, B, and C, in (53) we have

$$2N_3 = \frac{1}{2}\left(abc + bc + ca + ab + a + b + c + \frac{1}{3}\left(\frac{bc}{a} + \frac{ca}{b} + \frac{ab}{c}\right)\right) - 4$$

$$- 2\{s(bc,a) + s(ca,b) + s(ab,c)\}$$

$$- \frac{1}{4} + \frac{1}{6\,abc} + \frac{abc}{12} - \frac{1}{4}abc + \frac{1}{4}$$

$$= \tfrac{1}{3}abc + \tfrac{1}{2}(bc + ca + ab) + \tfrac{1}{2}(a + b + c) + \frac{1}{6}\left(\frac{bc}{a} + \frac{ca}{b} + \frac{ab}{c}\right) - 4 + \frac{1}{6abc}$$

$$- 2\{s(bc,a) + s(ca,b) + s(ab,c)\}.$$

The proof of Mordell's Theorem 5 is now complete.

It also is known (see [**46**] or Note 4) that, for a, b, and c pairwise coprime,

$$\left(s(bc,a) - \frac{bc}{12a}\right) + \left(s(ca,b) - \frac{ca}{12b}\right) + \left(s(ab,c) - \frac{ab}{12c}\right) \equiv -\frac{1}{4} - \frac{abc}{12} + \frac{1}{12abc} \pmod 2.$$

Using this, we obtain

$$N_3(a,b,c) \equiv \tfrac{1}{4}(a+1)(b+1)(c+1) \pmod 2.$$

It is easily seen that the corresponding formulae for lower dimensions

$$N_2(a,b) \equiv \tfrac{1}{2}(a+1)(b+1) \pmod 2,$$

$$N_1(a) \equiv (a+1) \pmod 2$$

are true. We are thus led to make the following

CONJECTURE. $N_4(a,b,c,d) \equiv \frac{1}{8}(a+1)(b+1)(c+1)(d+1)$ (mod 2).

Similar statements in higher dimensions may also be conjectured. That the present conjecture has some plausibility is shown by the fact that it holds at least for $d = 1$. Indeed,

$$N_4(a,b,c,1) = N_3(a,b,c),$$

since in $0 \leqq w < 1$ only $w = 0$ is possible.

Next,

$$\begin{aligned} N_3(a,b,c) &\equiv \tfrac{1}{4}(a+1)(b+1)(c+1) \\ &\equiv \tfrac{1}{8}(a+1)(b+1)(c+1)(1+1) \pmod 2, \end{aligned}$$

and this is precisely the content of the conjecture for $d = 1$. One can test also with some computation that the conjecture holds, e.g., for $N_4(2,3,5,7)$, $N_4(2,5,9,13)$, $N_4(3,4,5,7), \cdots$, but a complete proof does not seem to exist.

CHAPTER **4**

DEDEKIND SUMS AND MODULAR TRANSFORMATIONS

A. The η-Function and the Function $\Phi(M)$. The function $\eta(\tau)$ defined in Chapter **1** (see (3)) and now generally known as the Dedekind η-function, appears already in the work of Jacobi and Weierstrass on elliptic functions in the form

$$\Delta(\tau) = C\eta(\tau)^{24}. \tag{54}$$

Here C is an unimportant numerical constant. Actually, if one uses the proper normalization (see, e.g. [**31**], p. 108), then $C = 1$. This function has to do with the pattern of periods of the elliptic functions represented by the point lattice

$$\Omega = \{m_1\omega_1 + m_2\omega_2\}.$$

Here ω_1 and ω_2 are two generators of the point lattice and m_1, m_2 run independently through all integers. If ω_1', ω_2' is any other pair of generators of the same point lattice Ω, then

$$\begin{aligned} \omega_1' &= a\omega_1 + b\omega_2, \\ \omega_2' &= c\omega_1 + d\omega_2, \end{aligned} \tag{55}$$

where a, b, c, d are integers and

$$\begin{vmatrix} a & b \\ c & d \end{vmatrix} = 1. \tag{56}$$

A substitution (55) with property (56) is called a *modular substitution.* Instead of this homogeneous transformation we shall often use the *inhomogeneous* modular transformation, with

$$\tau = \frac{\omega_1}{\omega_2}, \qquad \tau' = \frac{\omega_1'}{\omega_2'},$$

so that instead of (55) we have $\tau' = (a\tau + b)/(c\tau + d)$. If we look at the lattice Ω first as $\Omega = \{\omega_1, \omega_2\}$ and then as $\Omega = \{\omega_1', \omega_2'\}$ it becomes clear that there should exist a relation between $\Delta(\tau)$ and $\Delta(\tau')$ and thus between $\eta(\tau)$ and $\eta(\tau')$, where obviously a certain 24th root of unity will have to be observed. In order to achieve this, Dedekind [**14**] does not consider $\eta(\tau)$, but

$$\log \eta(\tau) = \frac{\pi i \tau}{12} + \sum_{m=1}^{\infty} \log(1 - x^m), \; x = e^{2\pi i \tau}.$$

Here, since $\eta(\tau) \neq 0$ in the upper τ-half-plane, we can fix the meaning of the logarithm as

$$\log \eta(\tau) = \frac{\pi i \tau}{12} - \sum_{m=1}^{\infty} \sum_{r=1}^{\infty} \frac{1}{r} x^{mr}.$$

Under the modular transformation $\tau' = \tau + b$ we have simply

$$\log \eta(\tau + b) = \log \eta(\tau) + \frac{\pi i b}{12}. \tag{57a}$$

For $c > 0$, Dedekind proves now the important formula

$$\log \eta\left(\frac{a\tau + b}{c\tau + d}\right) \tag{57b}$$
$$= \log \eta(\tau) + \frac{1}{2}\log\left(\frac{c\tau + d}{i}\right) + \pi i\,\frac{a + d}{12c} - \pi i s(d, c),$$

where the principal branch is taken for the logarithm. Besides Dedekind's own proof there are several others in existence (see, e.g., [**61**] or [**25**]). We now observe that

$$\begin{pmatrix} a & b \\ c & d \end{pmatrix}\begin{pmatrix} 0 & -1 \\ 1 & 0 \end{pmatrix} = \begin{pmatrix} b & -a \\ d & -c \end{pmatrix}$$

and apply formula (57b) with $c > 0$, $d > 0$. With $\tau' = -1/\tau$ and $\tau'' = (a\tau' + b)/(c\tau' + d)$ we obtain successively that

$$\log \eta(\tau') = \log \eta(\tau) + \frac{1}{2}\log\frac{\tau}{i},$$

$$\log \eta(\tau'') = \log \eta(\tau') + \frac{1}{2}\log\left(\frac{c\tau' + d}{i}\right) + \pi i\,\frac{a + d}{12c} - \pi i s(d, c)$$

$$= \log \eta(\tau) + \frac{1}{2}\log\frac{\tau}{i} + \frac{1}{2}\log\left(\frac{-c/\tau + d}{i}\right) + \pi i\,\frac{a + d}{12c} - \pi i s(d, c)$$

$$= \log \eta(\tau) + \frac{1}{2}\log(c - d\tau) + \pi i\,\frac{a + d}{12c} - \pi i s(d, c).$$

On the other hand, using $\tau' = (b\tau - a)/(d\tau - c)$, we obtain

$$\log \eta(\tau'') = \log \eta(\tau) + \frac{1}{2}\log\left(\frac{d\tau - c}{i}\right) + \pi i\,\frac{b - c}{12d} - \pi i s(-c, d).$$

If we now equate the two expressions of $\log \eta(\tau'')$, we obtain

$$\pi i(s(d,c) - s(-c,d))$$

$$= \frac{\pi i}{12}\left(\frac{a+d}{c} - \frac{b-c}{d}\right) + \frac{1}{2}\log\left(\frac{c-d\tau}{d\tau-c}\,i\right)$$

or

$$s(d,c) - s(-c,d) = \frac{1}{12}\left(\frac{d}{c} + \frac{c}{d} + \frac{ad-bc}{dc}\right) + \frac{1}{2\pi i}\log(-i).$$

Here $s(-c,d) = -s(c,d)$ by (33a) and $ad - bc = 1$. Also (with our convention on logarithms) $\log(-i) = -\pi i/2$. With these remarks the last equality becomes

$$s(d,c) + s(c,d) = -\frac{1}{4} + \frac{1}{12}\left(\frac{d}{c} + \frac{1}{cd} + \frac{c}{d}\right)$$

and this is precisely the reciprocity formula (4) for Dedekind sums.

Dedekind used the sums $s(d,c)$ to describe the transformations of $\log\eta(\tau)$ under the substitutions $\tau' = (a\tau+b)/(c\tau+d)$. The integer c may be zero, positive, or negative. The case $c = 0$ is covered by formula (57a), and $c > 0$ by (57b), but so far we have not considered the case $c < 0$. In order to be able to treat all three cases together we introduce the symbol

$$\operatorname{sign} c = \begin{cases} 0 & \text{for } c = 0, \\ \dfrac{c}{|c|} & \text{for } c \neq 0. \end{cases}$$

The formula (57b) can be rewritten (still for $c \neq 0$), as

$$\text{(58)}\quad \log\eta\left(\frac{a\tau+b}{c\tau+d}\right) = \log\eta(\tau) + \tfrac{1}{2}(\operatorname{sign} c)^2\log\left(\frac{c\tau+d}{i(\operatorname{sign} c)}\right)$$

$$+ \pi i\,\frac{a+d}{12c} - \pi i \operatorname{sign} c\; s(d,|c|).$$

Indeed, if $c>0$, then sign $c=1, |c|=c$, and (58) reduces to (57b). If $c<0$, then $(a\tau+b)/(c\tau+d)=(-a\tau-b)/(-c\tau-d)$ with $-c>0$, so that

$$\eta\left(\frac{a\tau+b}{c\tau+d}\right)=\eta\left(\frac{-a\tau-b}{-c\tau-d}\right)$$

and (57b) is applicable. We now obtain, by using also (57b) and (33a), that

$$\eta\left(\frac{a\tau+b}{c\tau+d}\right)=\eta\left(\frac{-a\tau-b}{-c\tau-d}\right)=\eta(\tau)+\frac{1}{2}\log\left(\frac{-c\tau-d}{i}\right)$$

$$+\pi i\frac{-a-d}{-12c}-\pi i s(-d,-c)$$

$$=\eta(\tau)+\frac{1}{2}\log\frac{c\tau+d}{i\,\text{sign}\,c}+\pi i\,\frac{a+d}{12c}+\pi i s(d,|c|),$$

i.e., (58), because sign $c=-1$ and $(\text{sign}\,c)^2=1$. Let us now introduce a matrix function $\Phi(M)$, defined on the set of modular matrices $M=\begin{pmatrix}a & b\\ c & d\end{pmatrix}$ with a, b, c, and d rational integers and $ad-bc=1$. We define

$$(59)\quad \Phi(M)=\begin{cases} b/d & \text{for } c=0\\ \dfrac{a+d}{c}-12(\text{sign}\,c)s(d,|c|) & \text{for } c\neq 0.\end{cases}$$

One observes that

$$(60)\quad \log\eta\left(\frac{a\tau+b}{c\tau+d}\right)=\log\eta(\tau)$$

$$+\frac{1}{2}(\text{sign}\,c)^2\log\left(\frac{c\tau+d}{i\,\text{sign}\,c}\right)+\frac{\pi i}{12}\Phi\begin{pmatrix}a & b\\ c & d\end{pmatrix}$$

reduces to (57a) if $c = 0$ and to (57b) if $c \neq 0$, provided that the second term in (60) is understood to vanish if $\operatorname{sign} c = 0$.

The functional value $\Phi(M)$ is always an integer. Indeed, if $c = 0$, then $ad - bc = 1$ reduces to $ad = 1$, so that $a = d = \pm 1$, and, by (59), $\Phi(M) = b/d = \pm b$, an integer. If $c \neq 0$, then, again according to (59), $\Phi(M)$ is an integer provided that

$$a + d - 12|c|s(d,|c|)$$

is divisible by c. This is, indeed, the case and follows from the Reciprocity Theorem. We write the latter as

$$12h^2k\,s(h,k) + 12h^2k\,s(k,h) = -3h^2k + h^3 + hk^2 + h$$

and take congruences modulo k. Recalling that by Theorem 2 $2h^2s(k,h)$ is always an integer, we see that

$$12h^2k\,s(h,k) \equiv h^3 + h \pmod{k}.$$

Let h' be defined, by $hh' \equiv -1 \pmod{k}$; then $h^3 + h \equiv h^2(h - h') \pmod{k}$, and, using also the fact that $(h,k) = 1$, we obtain

$$12k\,s(h,k) \equiv h - h' \pmod{k}. \tag{61}$$

Since $ad - bc = 1$, it follows that $ad \equiv 1 \pmod{c}$, so that, if $k = |c|$ and $h = a$, one obtains $h' \equiv -d$. Substituting these in (61) leads to

$$12|c|s(a,|c|) \equiv a + d \pmod{|c|},$$

or by (33c) to $12|c|s(d,|c|) \equiv a + d \pmod{|c|}$ as claimed.

It is of group theoretical interest to study now the behavior of $\Phi(M)$ under composition of the modular substitutions.

We set as before $M = \begin{pmatrix} a & b \\ c & d \end{pmatrix}$, and also

$$M' = \begin{pmatrix} a' & b' \\ c' & d' \end{pmatrix} \text{ and } M'' = \begin{pmatrix} a'' & b'' \\ c'' & d'' \end{pmatrix}.$$

If $M'' = M'M$, i.e., if

$$\begin{pmatrix} a'' & b'' \\ c'' & d'' \end{pmatrix} = \begin{pmatrix} a' & b' \\ c' & d' \end{pmatrix} \begin{pmatrix} a & b \\ c & d \end{pmatrix},$$

then this matrix multiplication corresponds to the composition of the substitutions

$$\tau'' = \frac{a'\tau' + b'}{c'\tau' + d'}, \text{ and } \tau' = \frac{a\tau + b}{c\tau + d}.$$

The repeated application of (60) now leads to the result

$$\Phi(M'') = \Phi(M') + \Phi(M) - 3\,\text{sign}(c\,c'\,c''). \tag{62}$$

Indeed, by (60)

$$\log\eta(\tau') = \log\eta(\tau) + \tfrac{1}{2}(\text{sign}\,c)^2 \log\left(\frac{c\tau + d}{i\,\text{sign}\,c}\right) + \frac{\pi i}{12}\,\Phi(M)$$

and

$$\log\eta(\tau'') = \log\eta(\tau') + \tfrac{1}{2}(\text{sign}\,c')^2 \log\left(\frac{c'\tau' + d'}{i\,\text{sign}\,c'}\right) + \frac{\pi i}{12}\Phi(M'),$$

and adding these equations, we obtain

$$\log\eta(\tau'') = \log\eta(\tau) + \frac{1}{2}\left\{(\text{sign}\,c)^2 \log\left(\frac{c\tau + d}{i\,\text{sign}\,c}\right)\right.$$

$$\left. + \ (\text{sign}\,c')^2 \log\left(\frac{c'\tau' + d'}{i\,\text{sign}\,c'}\right)\right\} + \frac{\pi i}{12}\{\Phi(M) + \Phi(M')\}.$$

Also, if we apply (60) directly to $\tau'' = \dfrac{a''\tau + b''}{c''\tau + d''}$, we obtain

$$\log \eta(\tau'') = \log \eta(\tau) + \tfrac{1}{2}(\operatorname{sign} c'')^2 \log \frac{c''\tau + d''}{i \operatorname{sign} c''} + \frac{\pi i}{12}\Phi(M'').$$

From a comparison of these two values of $\log \eta(\tau'')$ it follows that

(62a) $$\Phi(M'') = \Phi(M) + \Phi(M') + R,$$

where

$$R = \frac{6}{\pi i}\left\{(\operatorname{sign} c)^2 \log\left(\frac{c\tau + d}{i \operatorname{sign} c}\right) + (\operatorname{sign} c')^2 \log\left(\frac{c'\tau + d'}{i \operatorname{sign} c'}\right)\right.$$
$$\left. - (\operatorname{sign} c'')^2 \log\left(\frac{c''\tau'' + d''}{i \operatorname{sign} c''}\right)\right\}.$$

If $c = c' = 0$, then $c'' = c'a + cd' = 0$, $R = 0$, and (62a) reduces to (62). If $c = 0$ and $c' \neq 0$, then $c'' = c'a$,

$$R = \frac{6}{\pi i}\log\left\{\frac{c'\tau' + d'}{c''\tau + d''}\operatorname{sign} a\right\}$$

and replacing c'' and $d'' = c'b + d'd$ by their values, we obtain

$$R = \frac{6}{\pi i}\log\frac{\operatorname{sign} a}{d} = 0,$$

because $c = 0$ forces $ad = 1$, so that $a = d = \pm 1$ and $(\operatorname{sign} a)/d = 1$. The cases $c' = 0$, $c \neq 0$ and $cc' \neq 0$, $c'' = 0$ are handled in the same way. Finally, if $cc'c'' \neq 0$ then

$$R = \frac{6}{\pi i}\log\left\{\frac{(c\tau + d)(c'\tau' + d')}{c''\tau + d''}\cdot\frac{\operatorname{sign} c''}{i(\operatorname{sign} c)(\operatorname{sign} c'')}\right\}.$$

If we replace here c'', d'' and τ' by their values, this simplifies to

$$R = -3 \cdot \frac{2i}{\pi} \log \frac{\operatorname{sign}(c''/cc')}{i} = -3 \operatorname{sign}(cc'c''),$$

and the proof of (62) is complete. We have proved (62) with the help of (60) by using the properties of the analytic function $\log \eta(\tau)$. But the definition (59) of Φ is a purely arithmetic one involving the Dedekind sums. The task, therefore, appears to prove (62) directly from the known properties of the Dedekind sums. This, indeed, has been done (see [**40**] and [**32**]), but we cannot reproduce here the lengthy proof.

B. An Application of Φ (M). The function Φ is useful in the study of the group structure of the modular group and its subgroups. As an example, let us consider the set Γ' of those modular substitutions M for which $\Phi(M) \equiv 0 \pmod 3$. It is clear from (62) that if $\Phi(M) \equiv \Phi(M') \equiv 0 \pmod 3$, then also $\Phi(M'') \equiv 0 \pmod 3$. Also, if $I = \begin{pmatrix} 1 & 0 \\ 0 & 1 \end{pmatrix}$ is the identity matrix, $\Phi(I) \equiv 0 \pmod 3$ and I belongs to Γ'. Finally, one verifies, using (62), that $\Phi(M) \equiv 0 \pmod 3$ implies that $\Phi(M^{-1}) \equiv 0 \pmod 3$ also holds. This shows that Γ' is a subgroup of index 3 of the full group of modular matrices. Γ' can be characterized as consisting of all substitutions, the matrices of which satisfy any one of the four congruences

$\begin{pmatrix} a & b \\ c & d \end{pmatrix} \equiv L \pmod 3$, where

$$L = \begin{pmatrix} 1 & 0 \\ 0 & 1 \end{pmatrix}, \begin{pmatrix} 0 & -1 \\ 1 & 0 \end{pmatrix}, \begin{pmatrix} 1 & 1 \\ 1 & -1 \end{pmatrix}, \text{ or } \begin{pmatrix} -1 & 1 \\ 1 & 1 \end{pmatrix}.$$

Here we remember that we don't distinguish between

$$\begin{pmatrix} a & b \\ c & d \end{pmatrix} \text{ and } \begin{pmatrix} -a & -b \\ -c & -d \end{pmatrix}$$

and that

$$\begin{pmatrix} a & b \\ c & d \end{pmatrix} \equiv \begin{pmatrix} a' & b' \\ c' & d' \end{pmatrix} \pmod n$$

means $a - a' \equiv b - b' \equiv c - c' \equiv d - d' \equiv 0 \pmod n$. The proof of the last statement reduces to the (very easy but lengthy) enumeration of all possible matrices (mod 3) and the elimination of those with $\Phi(M) \not\equiv 0 \pmod 3$.

C. The Class Invariant $\Psi(M)$. Two substitutions M_1, M_2 of the modular group Γ are called *similar*, if there exists an $L \in \Gamma$ such that

$$M_1 = L^{-1} M_2 L.$$

Similarity is an equivalence relation which divides the group Γ into *classes*. Direct computation shows that the trace $a + d$ is an invariant of the similarity classes. We obtain another invariant if we set

$$\Psi(M) = \Psi \begin{pmatrix} a & b \\ c & d \end{pmatrix} = \Phi(M) - 3 \operatorname{sign}(c(a + d)), \tag{63}$$

where $\Phi(M)$ is defined by (59). Since $\Phi(M)$ is an integer, the function $\Psi(M)$ is clearly also an integer, and we have

$$\Psi(M) = \Psi(-M),$$

$$\Psi(M^{-1}) = -\Psi(M).$$

Since

$$S = \begin{pmatrix} 1 & 1 \\ 0 & 1 \end{pmatrix} \text{ and } T = \begin{pmatrix} 0 & -1 \\ 1 & 0 \end{pmatrix}$$

generate the full group Γ (see, e.g., [**31**], p. 51 and 53), the invariance of Ψ will be proved if we show only that

$$\Psi(S^{-1}MS) = \Psi(M) \tag{64}$$

and

$$\Psi(T^{-1}MT) = \Psi(M). \tag{65}$$

Now, with $M = \begin{pmatrix} a & b \\ c & d \end{pmatrix}$ we have

$$S^{-1}MS = \begin{pmatrix} a - c & a - c + b - d \\ c & c + d \end{pmatrix} \tag{66a}$$

and

$$T^{-1}MT = \begin{pmatrix} d & -c \\ -b & a \end{pmatrix}. \tag{66b}$$

Equation (64) can now be inferred immediately from (66a), on account of the definitions (59) and (63).

For the proof of (65), we notice that

$$T^{-1} = \begin{pmatrix} 0 & 1 \\ -1 & 0 \end{pmatrix} \text{ and } T^{-1}M = \begin{pmatrix} c & d \\ -a & -b \end{pmatrix}.$$

Next, by (63) and (66b),

$$\Psi(T^{-1}MT) = \Phi(T^{-1}MT) - 3\,\text{sign}(-b(a + d)).$$

Now, according to the composition rule (62),

$$\begin{aligned} \Phi(T^{-1}MT) &= \Phi(T^{-1}M) + \Phi(T) - 3\text{ sign } ab \\ &= \Phi(T^{-1}) + \Phi(M) + \Phi(T) - 3\,\text{sign}\,ac - 3\,\text{sign}\,ab, \end{aligned}$$

so that

$$\Psi(T^{-1}MT) = \Phi(M) - 3\,\mathrm{sign}\,ac - 3\,\mathrm{sign}\,ab$$
$$+\ 3\,\mathrm{sign}\,(b(a+d)),$$

where we have taken into account the fact that $\Phi(T) = \Phi(T^{-1}) = 0$. By virtue of (63), our formula (65) is, therefore, equivalent to

$$\text{(67)}\quad \mathrm{sign}(c(a+d)) + \mathrm{sign}(b(a+d)) = \mathrm{sign}\,ac + \mathrm{sign}\,ab\,.$$

This equation is obviously correct for $d = 0$. Suppose, therefore, that $d > 0$. Then (67) still holds for $a \geqq 0$. For $a \leqq 0$, it follows that $ad \leqq 0$ and, hence, that $bc < 0$ and thus b and c are of opposite sign, and the equation holds again, as both sides vanish. The proof for $d < 0$ is entirely similar. We have thus derived the invariance property of the matrix function Ψ from the composition rule of the function Φ.

The invariant $\Psi(M)$ can also be expressed in such a way that it shows somewhat more about the structure of the modular group.

Put $U = ST$; then T and U are generators of the modular group Γ. This follows from the fact that $S = T^{-1}U$, and that, as we already recalled, S and T generate Γ. It also is easy to verify that

$$\text{(68)}\qquad T^2 = U^3 = I\,,$$

where we use the inhomogeneous notation, i.e., identify the matrices M and $-M$. Then any modular substitution can be written uniquely in shortest form as

$$M = U^{\varepsilon_0}TU^{\varepsilon_1}T\cdots TU^{\varepsilon_{\nu+1}}\,,$$

with $\varepsilon_j = \pm 1$, $j = 1, 2, \cdots, \nu$, while ε_0, $\varepsilon_{\nu+1} = 0$ or ± 1. If we remove a factor of M from the left of previous representation, and tag it on at the right end, or remove it from the right end, and tag it on at the left, then we say that we performed a "cyclic permutation" of the factors of M. If for example, the factor U is removed from the left and added on at the right, the result is $U^{-1}MU$. If the operation proceeds in the opposite direction, we obtain UMU^{-1}. In case the permuted factor is T, the corresponding results are $T^{-1}MT$ and TMT^{-1}, respectively. These cyclic permutations of the factors of M, being similarity transformations, will produce elements in the same class. Generally, proceeding step by step with cyclic permutations, we obtain reductions, until we arrive at a substitution in the same class and belonging to one of the following types:

(1) The elliptic T, U, U^{-1};

or

(2) $M = TU^{\varepsilon_1}TU^{\varepsilon_2}\cdots TU^{\varepsilon_\nu}$, $\varepsilon_j = \pm 1$.

It is clear that none of the matrices in (1) can be reduced further. It also is clear, on account of (68), that in (2) T can appear only to the first power and U only at powers $\varepsilon = \pm 1$. Indeed, $T^2 = I$, $U^3 = I$ and $U^2 = U^{-1}$. It may not be entirely obvious that we can always reduce M by similarity to a shortest form starting with T and ending with $U^{\pm 1}$. If M starts with a power of U and ends with T, a cyclic permutation brings it to the form (2) of same length. If M starts and ends with T, so that

$$M = TU^{\varepsilon_1}T\cdots TU^{\varepsilon_\nu}T, \tag{69}$$

say, a cyclic permutation leads, by (68), to $U^{\varepsilon_1}T\cdots TU^{\varepsilon_\nu}$, which has fewer factors than (69) and, by a further cyclic permutation, to $TU^{\varepsilon_2}T\cdots TU^{\varepsilon'}$, which is still shorter. Here

$$\varepsilon' = \varepsilon_\nu + \varepsilon_1 \equiv -1, 0, \text{ or } +1 \pmod 3).$$

If $\varepsilon' \equiv 0 \pmod 3$, then M is similar to

$$TU^{\varepsilon_2}T\cdots TU^{\varepsilon_{\nu-1}}T$$

of the same type as (69) and the process of shortening can be repeated. This procedure continues until either $\varepsilon' \equiv \pm 1$ (mod 3), so that one obtains a form of type (2), or else, until M is reduced to a single factor, which is necessarily one of those listed under (1). For the matrices of type (1), one immediately verifies, using (63), that

$$\Psi(T) = 0, \Psi(U) = \Psi\begin{pmatrix}1 & -1\\ 1 & 0\end{pmatrix} = -2, \text{ and } \Psi(U^{-1}) = 2.$$

For the matrices of type (2), the following lemma holds.

LEMMA 7. *If the modular matrix M is similar to one of type* (2), *say,*

$$M \sim TU^{\varepsilon_1}\cdots TU^{\varepsilon_\nu},$$

then

$$\Psi(M) = \sum_{j=1}^{\nu} \varepsilon_j. \tag{70}$$

In order to prove (70), we need the auxiliary

LEMMA 8 (see [**49**]). *If*

$$M = \begin{pmatrix}a & b\\ c & d\end{pmatrix} = TU^{\varepsilon_1}T\cdots TU^{\varepsilon_\nu}, \ \varepsilon_j = \pm 1, \ j = 1, 2, \cdots, \nu,$$

then

$$a < 0\,, d < 0\,, b \geqq 0\,, c \geqq 0 \tag{71}$$

hold either for M or for $-M$.

Proof of Lemma 8. First we have

$$TU = \begin{pmatrix} -1 & 0 \\ 1 & -1 \end{pmatrix},\ TU^{-1} = \begin{pmatrix} 1 & -1 \\ 0 & 1 \end{pmatrix},$$

$$\Psi(TU) = \Psi(TST) = \Psi(S) = \Psi\begin{pmatrix} 1 & 1 \\ 0 & 1 \end{pmatrix} = 1\,,$$

and

$$\Psi(TU^{-1}) = \Psi(TT^{-1}S^{-1}) = \Psi(S^{-1}) = \Psi\begin{pmatrix} 1 & -1 \\ 0 & 1 \end{pmatrix} = -1,$$

which show that (70) and (71) are both correct for $\nu = 1$. To finish the proof of the lemma by induction, let $\varepsilon_{\nu+1} = 1$, $M_1 = MTU^{\varepsilon_{\nu+1}} = MTU$, and observe again that

$$TU = \begin{pmatrix} -1 & 0 \\ 1 & -1 \end{pmatrix}.$$

Hence, if $M = \begin{pmatrix} a & b \\ c & d \end{pmatrix}$ satisfies (71), then

$$MTU = \begin{pmatrix} a & b \\ c & d \end{pmatrix}\begin{pmatrix} -1 & 0 \\ 1 & -1 \end{pmatrix} = \begin{pmatrix} -a+b & -b \\ -c+d & -d \end{pmatrix}$$

$$= \begin{pmatrix} a-b & b \\ c-d & d \end{pmatrix} = \begin{pmatrix} a_1 & b_1 \\ c_1 & d_1 \end{pmatrix}$$

and a_1, b_1, c_1 and d_1, again satisfy (71). The next to last equality is justified because we identify any matrix with its negative. The case $\varepsilon_{\nu+1} = -1$ is handled in the same way and Lemma 8 is proved.

Proof of Lemma 7. For a proof by induction of (70), let again $M_1 = MTU^{\varepsilon_{\nu+1}}$. We have to show that

$$\Psi(MTU^{\varepsilon_{\nu+1}}) = \Psi(M) + \varepsilon_{\nu+1}.$$

For $\varepsilon_{\nu+1} = 1$, on the one hand, by (63),

$$\begin{aligned}\Psi(MTU) &= \Phi(MTU) - 3\,\mathrm{sign}((-c+d)(-a+b-d))\\ &= \Phi(MTU) + 3,\end{aligned}$$

in view of (71). Moreover,

$$\begin{aligned}\Phi(MTU) &= \Phi(M) + \Phi(TU) - 3\,\mathrm{sign}(c(-c+d))\\ &= \Phi(M) - 2 + 3\,\mathrm{sign}\, c\end{aligned}$$

by virtue of (59) and (71), so that

$$\Psi(MTU) = \Phi(M) + 1 + 3\,\mathrm{sign}\, c.$$

On the other hand we infer from (63), taking again into account (71), that $\Psi(M) = \Phi(M) + 3\,\mathrm{sign}\, c$, so that we arrive at the desired result

$$\Psi(MTU) = \Psi(M) + 1.$$

The case $\varepsilon_{\nu+1} = -1$ can be handled in the same manner.

The similarity classes of modular substitutions are in one to one correspondence with the proper equivalence classes of binary quadratic forms of discriminant $\Delta = m^2 - 4$, where m is the trace of the modular matrix. The above discussed property of Ψ, i.e., the fact that it is a class invariant for similarity classes of modular matrices leads then to

THEOREM 6. *For the quadratic forms* $Ax^2 + Bxy + Cy^2$ *of discriminant* $\Delta = B^2 - 4AC = m^2 - 4$, *the function*

$$\Psi\begin{pmatrix} \frac{m-B}{2} & -C \\ A & \frac{m+B}{2} \end{pmatrix}$$

is a class invariant.

If in this theorem one takes particular, small values for m, several theorems about $s(h,k)$ follow, of which we quote only one as an example.

THEOREM 7. *If* $k > 0$ *and* $(h-2)^2 \equiv 3 \pmod k$, *then*

$$s(h,k) = \begin{cases} \dfrac{1-k}{3k} & \text{for } k \equiv 1 \pmod 3 \text{ and for } k \equiv -3 \pmod 9 \\[2ex] \dfrac{2-k}{6k} & \text{for } k \equiv -1 \pmod 3 \text{ and for } k \equiv 3 \pmod 9). \end{cases}$$

For a proof see [**49**].

D. The Dedekind Sums and the Theory of Partitions. The Dedekind sums also have applications in the theory of partitions. If $p(n)$ is the number of unrestricted partitions of n, then we know since Euler that

$$\sum_{n=0}^{\infty} p(n)x^n = \frac{1}{\prod_{m=1}^{\infty}(1-x^m)},$$

where $p(0) = 1$. If one sets $x = e^{2\pi i\tau}$, then the denominator on the right side is essentially (i.e., except for a factor $e^{\pi i\tau/12}$) identical with $\eta(\tau)$. From (57b) we derive the transformation formula of $\eta(\tau)$ by exponentiation. For $c > 0$,

$$\eta\left(\frac{a\tau + b}{c\tau + d}\right) = \varepsilon(a, b, c, d,) \sqrt{\frac{c\tau + d}{i}}\eta(\tau),$$

where

$$\varepsilon(a, b, c, d) = \exp\left\{\pi i\left(\frac{a + d}{12c} - s(d, c)\right)\right\},$$

which is a root of unity. It is clear that in the theory of transformation of $\eta(\tau)$, the Dedekind sum enters only through its values modulo 2. The study of $p(n)$ using the theory of $\eta(\tau)$ was inaugurated by the famous investigations of Hardy and Ramanujan and has led through some further refinements to the following results [**52**]:

$$p(n) = \frac{1}{\pi\sqrt{2}} \sum_{k=1}^{\infty} A_k(n) k^{1/2} \frac{d}{dn}\left(\frac{\sinh(C\lambda_n/k)}{\lambda_n}\right),$$

with the abbreviations

$$C = \pi\sqrt{\frac{2}{3}} \qquad \lambda_n = \sqrt{n - \frac{1}{24}}\ , \quad \text{and}$$

$$A_k(n) = \sum_{\substack{h \bmod k \\ (h,k)=1}} \omega_{hk} e^{-2\pi i hn/k}\ , \quad \omega_{hk} = e^{\pi i s(h,k)}\ .$$

The function $\eta(\tau)$ is actually a modular form of dimension $-1/2$. Modular forms of other real dimensions can be defined, and in their theory a fuller knowledge of the Dedekind sums is essential.

E. Class Number Formulae. Another appearance of the Dedekind sums is worth mentioning, namely in certain class number formulae of abelian fields over quadratic ground fields. The theory was outlined by Hecke and was carried out in detail by C. Meyer [**33**]. The formulae and their background are too complex to be quoted and explained here.

CHAPTER **5**

GENERALIZATIONS

We shall close this presentation of Dedekind sums with the mention of some of their generalizations. There are several in which the function $((x))$, which is essentially the first Bernoulli polynomial $B_1(y) = y - \frac{1}{2}$ of $y = x - [x]$, is replaced by higher Bernoulli polynomials. They play roles in special problems of partitions.

In the work of C. Meyer [**32**], [**33**], but also already in some investigations by J. Lehner [**29**] and J. Livingood [**30**], certain Dedekind sums appear in which the summand μ is restricted by congruence conditions. These types of Dedekind sums are in full generality contained in the definition

$$s(h,k;x,y) = \sum_{\mu \bmod k} \left(\left(\frac{\mu+y}{k}\right)\right)\left(\left(h\frac{\mu+y}{k}+x\right)\right),$$

where x and y are real numbers. Such a sum is clearly of period 1 in x and in y. This sum possesses the following reciprocity formula

$$s(h,k;x,y) + s(k,h;y,x) = -\tfrac{1}{4}\delta(x)\delta(y) + ((x))((y))$$

$$+ \frac{1}{2}\left\{\frac{h}{k}\Psi_2(y) + \frac{1}{hk}\Psi_2(hy+kx) + \frac{k}{h}\Psi_2(x)\right\},$$

where $\Psi_2(x) = B_2(x - [x])$. Here $B_2(x)$ is the second Bernoulli polynomial and

$$\delta(x) = \begin{cases} 1 & \text{if } x \text{ is an integer,} \\ 0 & \text{otherwise.} \end{cases}$$

These generalized Dedekind sums have some interesting properties, which are discussed in [**50**].

Let us finish by mentioning one more problem, one that is not yet solved. What are the expressions corresponding to the Dedekind sums in algebraic fields? There seem to be two ways to study this question. Firstly, one might ask what does the function $\eta(\tau)$ become in algebraic fields? Such a question has, of course, no unique answer. But Hecke ([**24**], p. 202) has indicated a function belonging to real quadratic fields, which shares many properties with $\log \eta(\tau)$. The transformation formula of that Hecke function has not been worked out in detail yet and Hecke gave only the special case which corresponds to $\tau' = -1/\tau$. This should be a rewarding problem.

Another way would be to simulate the function $[x]$ in algebraic fields. Now $[x]$ implies an order among the numbers x, which we may assume to be rational. So one would have to look around for a natural order of the numbers of an algebraic field. Such an approach has been made by G. J. Rieger [**56**]. As of now, there is no proof available that the Dedekind sums defined in these ways share with the ordinary Dedekind sums some of the latter's important properties.

CHAPTER **6**

SOME REMARKS ON THE HISTORY OF THE DEDEKIND SUMS

Bernhard Riemann died on July 20, 1866 at the age of forty. According to his wish, his manuscripts, notes, etc., were entrusted to R. Dedekind. It turned out that this was a rather mixed lot; there were some practically finished papers, then some drafts of varying degrees of completeness, and some that were just fragmentary sketches.

Among the latter were two notes related to the theory of elliptic modular function as presented in Jacobi's *Fundamenta Nova*. In Jacobi's work, the parameter q satisfies $|q| < 1$, while in these notes, Riemann considers the limiting case $|q| = 1$. The first note contains 68 formulae and is written in Latin. The second one consists of a single sheet of paper, is written almost illegibly, and contains eight formulae without any text (except for the three qualifiers "gerade" (even), "ungerade" (odd), and "absolut kleinster Rest von x" (residue of least absolute value of x), all in German).

Dedekind felt unable to edit himself all of Riemann's unfinished papers and asked and obtained the cooperation, first of A. Clebsch (in 1872) and, after Clebsch's untimely death, that of H. Weber (in 1874). Riemann's collected papers, including his posthumous work, came out in 1876 with H. Weber as general editor; however, the two notes dealing with Jacobi's elliptic functions were edited by R. Dedekind himself, under the title *Fragmente über Grenzfälle der elliptischen Modulfunktionen.* Following this paper, Dedekind published a set of comments containing:

(a) Some historical remarks and conjectures (such as Dedekind's opinion that at least the first note had been written already in 1851, i.e., 14 or 15 years before Riemann's death);

(b) some explanations which may facilitate the reading of Riemann's fragments; and

(c) an application of Riemann's method to a problem not actually occurring in Riemann's two notes.

This problem, already considered by Jacobi and Hermite, is the study of the behavior of the function

$$\eta(\omega) = q^{1/12} \prod_{\nu=1}^{\infty} (1-q^{2\nu}), \quad q = e^{\pi i \omega}, \omega = x + iy, \quad y>0,$$

when ω is subjected to linear fractional transformations. This addendum of Dedekind, entitled *Erläuterungen zu den Fragmenten XXVIII*, has become justly famous, and the function $\eta(\omega)$ is now generally known as Dedekind's η-function.

If α, β, γ, and δ are integers, $\alpha\delta - \beta\gamma = 1$ and $\omega'' = \alpha + \beta\omega, \omega' = (\gamma + \delta\omega)/\omega''$, then Dedekind proves that

$$\log \eta(\omega') = \log \eta(\omega) + \frac{1}{2} \log \frac{\omega''}{i} + \frac{\pi i}{12} S,$$

where $S = S(\alpha, \beta, \gamma, \delta)$ is an integer, uniquely determined by α, β, γ, and δ. After a short statement concerning the relevance of this integer, the balance of Dedekind's paper is devoted to the complete determination of S and to its representation by a finite sum—essentially (i.e., except for a change in notation) the Dedekind sum. In fact, a comparison with (57b) immediately shows that

$$S(\alpha,\beta,\gamma,\delta) = \frac{\beta+\gamma}{12\alpha} - s(\beta,\alpha).$$

While $S(\alpha,\beta,\gamma,\delta)$ is an integer, $s(\beta,\alpha)$, in general, is not integral, but may have as its denominator any divisor of 6α (see Chapter 3(A), Theorem 2—all indications of chapters and pages refer to the present book).

Dedekind also mentions that Riemann's eight formulae of Part II are a consequence of certain results (involving Dedekind sums) obtained in the "*Erläuterungen*", but the connection is non-trivial and was never presented in detail by Dedekind. In fact, the proof of Riemann's formulae was given for the first time by Rademacher and Whiteman [**44**].

As stated in Chapters 1 and 4, the methods used by Dedekind are of a transcendental nature and make essential use of the work of Jacobi and Riemann on elliptical modular functions. The result, however, is of an elementary nature: an integer, represented as a finite sum of fractions. It also seems clear that it has arithmetic significance, and it appeared eminently desirable to study the Dedekind sums by elementary methods. This was accomplished by H. Rademacher in a series of some ten papers, starting in 1931 (see the bibliography).

The keystone of this theory of Dedekind sums is their reciprocity law (see (4) in Chapter 1 and Theorem 1 in Chapter 2). As mentioned in Chapter 1, a first proof of this theorem (by a transcendental method) is due already to Dedekind himself.

At present there exist quite a few essentially distinct proofs of the reciprocity law, most of them due to Rademacher, and some are quite elementary. In Chapter 2 one finds a representative selection (but by no means all) of these proofs.

Many other properties of the Dedekind sums, especially those of an arithmetic nature, were investigated by Rademacher, and some of the results obtained are found in Chapters 3 and 4. Among the most important ones not included there, are some obtained in collaboration with A. L. Whiteman [**44**], where one finds, as already mentioned earlier, the first (and presumably only) published proof of the formulae stated (without proof) by Riemann in the second of his "*Fragmente*..." [**57**].

Rademacher's work on Dedekind sums found many important applications and also sparked the interest of other mathematicians in the study of this topic. A very brief, roughly chronological, but forcibly incomplete survey of these contemporary developments follows and concludes this historical sketch.

In 1950 and 1952, T. M. Apostol ([**1**] and [**2**]) generalized the definition of Dedekind sums as follows:

Let $B_n(x)$ be the nth Bernoulli polynomial and set $\bar{B}_n(x) = B_n(x - [x])$; then, for non-integral x, $\bar{B}_1(x) = ((x))$, and for $(h, k) = 1$ the Dedekind sum may be written as

$$s(h,k) = \sum_{\mu(\mathrm{mod}\, k)}{}' \bar{B}_1\left(\frac{\mu}{k}\right)\bar{B}_1\left(\frac{\mu h}{k}\right).$$

Here and in what follows, the prime stands for the restriction $\mu \not\equiv 0 \pmod{k}$, and the $\bar{B}_n(x)$ will be called *Bernoulli functions*.

For integral p, Apostol defined the closely related sums

$$s_p(h,k) = \sum_{\mu=1}^{k-1} \frac{\mu}{k} \bar{B}_p\left(\frac{h\mu}{k}\right),$$

and used them to obtain transformation formulae for functions represented by Lambert series

$$G_p(x) = \sum_{n=1}^{\infty} n^{-p}x^n/(1-x^n).$$

He also studied the relation of these sums to the Hurwitz zeta function $\zeta(s,a)$.

In addition to these results, T. Apostol obtained in his dissertation (Berkeley, 1948) several others, that were never published. Among these are some found also by Rademacher and published in [**49**], such as

(i) $s(h,k) = 0$ if and only if $h^2+1 \equiv 0 \pmod{k}$,

and

(ii) $12hk\,s(h,k) = (k-1)(k-h^2-1)$ if $k \equiv 1 \pmod{h}$.

It is clear that (i) follows from the result of Chapter 3A (see p. 28; see also [**49**]) which states that $12s(h,k)$ cannot be an integer unless $h^2+1 \equiv 0 \pmod{k}$, when $s(h,k) = 0$. Other interesting results of Apostol, giving explicit values for $s(h,k)$, such as

(iii) $12\,hk\,s(h,k) = (k-2)\left(k - \dfrac{h^2+1}{2}\right)$ if $k \equiv 2 \pmod{h}$,

and

(iv) $12\,hk\,s(h,k) = k^2 + (h^2-6h+2)k + h^2 + 1$

if $$k \equiv -1 \pmod{h}$$

have apparently never been published.

In 1933 L. Rédei [**54**] gave an elementary proof of the reciprocity formula (4). In 1950 he obtained [**55**] a new proof and showed that this is only one of an infinite sequence of similar relations.

In fact, let $(m, n) = 1$ and for any integer k set $F_k(x) = (x^k - 1)/(x - 1)$. Then

$$F_m(x)X_{mn}(x) + F_n(x)X_{nm}(x) = 1$$

has polynomial solutions $X_{mn}(x)$ and $X_{nm}(x)$ of maximal degrees $n-1$ and $m-1$, respectively. These solutions can be obtained explicitly. If one now replaces in the previous equation x by $1 + t$, then the vanishing of the coefficients of t^k for $k = 1, 2, \cdots$ leads to a sequence of identities. For $k = 2$, in particular, one obtains the reciprocity formula (4) of the Dedekind sums.

In 1951, L. J. Mordell published the papers [**36**] and [**37**]. In [**36**] he studied sums of the form $\Sigma_K f(qx + py)$, taken over the lattice points of a region K, defined by $0 < x < p$, $0 < y < q$, $0 < qx + py < pq$. He thus obtained reciprocity relations for these generalized sums. The case $f(u) = u$ corresponds to the classical Dedekind sums. In [**37**] he studied by elementary means the relations between the lattice points in a tetrahedron and the Dedekind sums. Some of his results are incorporated in Chapter 3 (E).

L. Carlitz wrote several papers on Dedekind sums and their generalizations (see the bibliography), the first paper [**5**] appearing in 1953. For fixed odd $p > 1$, $(h, k) = 1$, and $0 \leqq r \leqq p + 1$, he considered the sums (already occurring in Apostol's work [**1**])

$$c_r(h, k) = \sum_{\mu \pmod k} \bar{B}_{p+1-r}\left(\frac{\mu}{k}\right) \bar{B}_r\left(\frac{h\mu}{k}\right).$$

He studied their properties and their relations to Bernoulli and Euler numbers, deduced from Apostol's transformation formula a reciprocity relation for the sums $c_r(h,k)$ (i.e., a relation between these sums and those obtained by interchanging h and k), obtained Apostol's formula as a particular case of a more general result, and indicated a new proof (see Chapter 2 (C_1)) for the ordinary and generalized Dedekind sums.

Carlitz also generalized the Dedekind sums in a new way as follows: Let

$$f\left(\frac{r}{k}\right) = \left(\left(\frac{r}{k}\right)\right) + \frac{1}{2k}$$

and set

$$s_n(h_1, \cdots, h_n; k) = \sum_{r_1, \cdots, r_n (\mathrm{mod}\, k)} f\left(\frac{r_1}{k}\right) \cdots f\left(\frac{r_n}{k}\right) f\left(\frac{r_1 h_1 + \cdots + r_n h_n}{k}\right).$$

One observes that for $n = 1$, s_1 is essentially a Dedekind sum. Relations involving either $n+1$, or $n+2$ terms follow for the s_n; for $n = 1$, these reduce to the reciprocity formula (4) and to a 3-term relation of Rademacher [**47**], respectively. Also the relation of the s_n to the higher Bernoulli numbers is elucidated. In more recent papers ([**11**], [**12**], [**13**]), Carlitz generalized the sums $s(h,k;x,y)$ and proved reciprocity formulae and three-term relations for these sums.

In 1957, C. Meyer [**32**] generalized the Dedekind sums for arguments of arbitrary sign and indicated their reciprocity relation, by interpreting them as

$$\sum_{k \bmod n}' \bar{B}_1\left(\frac{mk}{n}\right) \bar{B}_1\left(\frac{k}{n}\right).$$

He considered similar sums over Bernoulli functions of higher degree and obtained many interesting formulae.

A generalization of the function $\Phi(M)$ is shown to be an invariant of the ring classes of a real quadratic field.

Meyer also introduced the following new generalization (this has been generalized further by Rademacher; see Chapter 5) of the Dedekind sums

$$s_{gh}(a,c) = \sum_{\mu \bmod c}{}' \; \bar{B}_1\left(\frac{a\mu}{c} + \frac{ag+ch}{fc}\right)\bar{B}_1\left(\frac{\mu}{c} + \frac{g}{fc}\right),$$

and the corresponding generalization of $\Phi(M)$. Finally, he used these generalizations in order to obtain the transformation formulae for Klein's functions

$$\sigma_{gh}(\omega_1,\omega_2) = \sigma(u,\omega_1,\omega_2),$$

where $\sigma(u,\omega_1,\omega_2)$ is Weierstrass' σ-function, f,g, and h are rational integers, and $u = (g\omega_1 + h\omega_2)/f$.

In [**32**a] Meyer obtained the generalization of Rademacher's cotangent formulas (25), (26) for the generalized Dedekind sums $s_{gh}(a,c)$.

In his monograph [**33**], Meyer obtained an explicit Kronecker type limit formula (involving Dedekind sums) for L-functions of ray classes in real quadratic fields and used these results for the determination of the class number of abelian extensions of quadratic fields.

U. Dieter [**16**], [**17**], [**18**], a former student of Rademacher, gave a new neat proof of the reciprocity law for Dedekind sums (see first proof in Chapter 2), and obtained additional results (some new, some already known) concerning sums of three Dedekind sums $\sum_{i=1}^{3} s(d_i,c_i)$, where (d_i,c_i) are the second rows of matrices whose product equals $\begin{pmatrix} 0 & 0 \\ 1 & 1 \end{pmatrix}$.

Furthermore he extended results of C. Meyer by determining the behavior of Klein's function $\log \sigma_{g,h}(\omega_1, \omega_2)$ under arbitrary transformations of the modular group. He obtained the reciprocity formula (see Chapter 5) for Meyer's generalized Dedekind sums $s_{g,h}(a,c)$ both by analytic and by arithmetic methods.

Later (see [**19**], [**20**], [**21**]) Dieter discovered a strong connection between generalized Dedekind sums and PRN (Pseudo-Random-Numbers). These numbers are generated by Lehmer's linear congruential method: let m, a, r, y_0 be integers and let y_i be defined by

$$y_{i+1} \equiv ay_i + r (\operatorname{mod} m), \; 0 \leqq y_i < m.$$

The fractions $x_i = y_i/m$ are the PRN. Dieter showed that the probability distribution of pairs of PRN, the frequency of permutations of three PRN and the serial correlation between x_i and x_{i+s} can be calculated by means of generalized Dedekind sums. It is interesting to remark that this rather surprizing connection between the serial correlation and Dedekind sums was discovered independently and roughly simultaneously also by B. Jansson ([**26**], especially Chapter 5) and D. E. Knuth ([**27**]; see also p. 78).

In 1957 two papers by M. Mikolás ([**34**] and [**35**]) appeared with the following generalizations of Dedekind sums:

Let $\{u\} = u - [u]$ and for m, n non-negative integers and $(a,c) = (b,c) = 1$, set

$$\mathfrak{s}\begin{pmatrix} a & b \\ & c \end{pmatrix} = \sum_{\lambda=0}^{c-1} \bar{B}_m\left(\frac{\lambda a}{c}\right) \bar{B}_n\left(\frac{\lambda b}{c}\right),$$

and

$$S_c^{a,b}(x,y) = \sum_{\lambda (\operatorname{mod} c)} \exp\left\{2\pi i\left(\left\{\frac{\lambda a}{c}\right\}x + \left\{\frac{\lambda b}{c}\right\}y\right)\right\}.$$

Finally, for nonintegral x and y, set

$$\mathfrak{S}_c^{a,b}(x,y) = (e^{2\pi i x} - 1)^{-1}(e^{2\pi i y} - 1)^{-1} S_c^{a,b}(x,y).$$

These functions turn out to generalize the Dedekind sums, although this is hardly obvious.

In [34] Mikolás established a large number of beautiful identities involving these functions, e.g.,

$$\mathfrak{S}_a^{c,b}(ax + by, -cy) - \mathfrak{S}_b^{c,-a}(ax + by, cx) + \mathfrak{S}_c^{a,b}(cx, cy) = 0.$$

One may now use the connection between the zeta functions and the Bernoulli functions, to show that the sums

$$\mathfrak{D}_c^{a,b}(\omega, z) = \sum_{\lambda=1}^{c-1} \zeta\left(\omega, \left\{\frac{\lambda a}{c}\right\}\right)\zeta\left(z, \left\{\frac{\lambda b}{c}\right\}\right)$$

($(a,c) = (b,c) = 1$, $c > 1$, $\zeta(s,a) =$ Hurwitz' zeta function) are further generalizations of the Dedekind sums. Mikolás studied these $\mathfrak{D}_c^{a,b}$ in detail and determined the functional equations that they satisfy.

In [35] the function

$$\tilde{Q}(\tau, \omega) = \sum_{n=1}^{\infty} \left(\frac{1}{n + \omega} + \frac{1}{n - \omega}\right) \frac{e^{2\pi i n \tau}}{1 - e^{2\pi i n \tau}}$$

is investigated following essentially the method of [48]. The transformation formula for $\tilde{Q}(\tau, \omega)$ under modular transformations of the variables involves the functions $\mathfrak{S}_a^{b,c}(x,y)$. The very general results obtained contain most of the known identities concerning Dedekind sums, including the reciprocity laws.

In 1959, K. Wohlfahrt [63] made use of the way in which Dedekind sums appear in the multipliers of certain modular forms, in order to construct subgroups of the modular group. Incidentally, he proved that no general divisibility properties may be expected for Dedekind sums, except for some known congruences modulo 24.

In 1960, G. J. Rieger [**56**] extended the concept of Dedekind sums to algebraic number fields (see Chapter 5, p. 65). The starting point of Rieger is again the representation

$$s(h,k) = \sum_{\mu(\bmod k)}{}' \bar{B}_1\left(\frac{\mu}{k}\right)\bar{B}_1\left(\frac{\mu h}{k}\right), \tag{1'}$$

for $(h,k) = 1$ and with $\bar{B}_1(x)$ the Bernoulli function defined on p. 69. If one sets $x = h/k$ with $(h,k) = 1$ and defines $D(x) = s(h,k)$, then one may consider $D(x)$ as a Dedekind sum defined on the rationals*. The purpose of Rieger was to extend this definition from rational x to arbitrary algebraic x. In order to achieve this, Rieger considered Eisenstein's formula for the Bernoulli function $\bar{B}_1(x)$, when $x = u/v$ is rational:

$$\bar{B}_1\left(\frac{u}{v}\right) = \frac{i}{2v}\sum_{r=1}^{v-1} e^{2\pi i r u/v} \cot\frac{\pi r}{v}.$$

He first generalized this formula to the case of $x \in K$, where K is a finite algebraic extension of the rationals. Next, he substituted these generalized Bernoulli functions in a formula similar to (1′) and obtained in this way the generalization $D_K(v)$ (v an algebraic integer in K) of $D(x)$, from the rational field to any field K, algebraic over the rationals. The details of this work are quite technical. For this reason,

* Actually, one may verify by means of (1), that if $m = th$, $n = tk$ with $(h, k) = 1$, then $s(m, n) = s(h, k)$ and $D(x)$ is well defined for rational $x = u/v$ even without the restriction $(u, v) = 1$.

we shall content ourselves here with only two remarks. The first should convey something of the flavor of Rieger's work, while the second will suggest its relevance.

The role of h and k is now played by the integral ideals $\mathfrak{c}$ and $\mathfrak{b}$ in K. Let $\mathfrak{o}$ and $\mathfrak{d}$ stand for the unit ideal of integers and for the different of K, respectively. The trivial generalization of $x = u/v$ would be the ideal $\mathfrak{c}\mathfrak{b}^{-1}$, but even a modest familiarity with algebraic numbers will suggest to consider instead the ideal $\mathfrak{c}\mathfrak{b}^{-1}\mathfrak{d}^{-1}$. Now consider the set $\mathfrak{c}\mathfrak{b}^{-1}\mathfrak{d}^{-1} \cap \mathfrak{o}$ of elements of K. It turns out that for v an algebraic integer in K the Dedekind sums $D_K(v)$ take on the same value for all $v \in \mathfrak{c}\mathfrak{b}^{-1}\mathfrak{d}^{-1} \cap \mathfrak{o}$.

The second remark concerning the $D_K(v)$ is that if K reduces to the rational field, then the $D_K(v)$ are rational and reduce to the ordinary Dedekind sums. This shows that the $D_K(v)$ are genuine, nontrivial extensions of the classical Dedekind sums, from the rational field to arbitrary finite algebraic extension fields of the rationals.

In 1969 K. Barner [4] defined

$$S_{2n}^{(m)}(\delta,\gamma) = \sum_{\mu(\mathrm{mod}\ \gamma)} \bar{B}_m\left(\frac{\delta\mu}{\gamma}\right)\bar{B}_{2n-m}\left(\frac{\mu}{\gamma}\right)$$

for natural integers n and m with $0 \leqq m \leqq 2n$, and where γ, δ, μ are rational integers, $\gamma > 0$, $(\delta,\gamma) = 1$ and $\bar{B}_m(x)$ are the Bernoulli functions defined on p. 69. He used this generalization of the Dedekind sums in order to give explicit formulae for the values at integral arguments s, of zeta and L-functions corresponding to ring classes in real quadratic number fields. It turns out that (if the parity of s is right in relation to the defining character of the L-function) these values are of the form $r\pi^{2s}\sqrt{d}$, where d is the discriminant of the field and r is rational.

As a fairly recent and possibly somewhat unexpected contribution to the theory and applications of Dedekind sums, let us consider again the work of Donald E. Knuth, already mentioned in connection with related work of U. Dieter. In 1969, in the second volume of his treatise on *The Art of Computer Programming* [**27**], Knuth used a slight generalization of the classical Dedekind sums, namely

$$\sigma(h,k,c) = 12 \sum_{0 \leqq j \leqq k} \left(\left(\frac{j}{k}\right)\right) \left(\left(\frac{hj+c}{k}\right)\right),$$

in his study of statistical (more specifically, of serial correlation) tests. He reduced the proof of the reciprocity formula for $\sigma(h,k,c)$ to that for $\sigma(h,k,0)$, i.e., essentially for the classical Dedekind sum, and proved it in this case by a streamlined version of Carlitz's proof (see Chapter 2(C_1) and Note 1). He also indicated an efficient method for the numerical computation of $\sigma(h,k,c)$.

We shall conclude this brief survey with the mention of work by B. Schoeneberg. In 1967 (see [**62**]) he considered (in a notation different from that of Meyer and Dieter) the transformation formula of $\log \sigma_{gh}(\omega_1, \omega_2)$, ($\sigma_{gh}(\omega_1,\omega_2)$ is Klein's function and is defined on p. 73). He observed that $\log \sigma_{gh}(\omega_1,\omega_2)$ is an integral of the third kind and used this remark thoroughly. Generalized Dedekind sums appear in the transformation formula.

Much work on Dedekind sums, their generalizations and their applications is still in progress. But the time to report on it has not yet come.

Added in proof. In 1971 F. Hirzebruch published several papers in which he arrived at the Dedekind sums by starting from purely topological considerations. Using his signature theorem and deep results of Attiyah-Bott-Singer on group actions on 4-dimensional manifolds, he obtained various results on Dedekind sums: a three term formula of Rademacher (see [**47**]) from which the reciprocity formula (4) follows quite trivially, Dedekind's formula (42), and Mordell's Theorem 5. D. Zagier has written a paper on "Equivariant Pontrjagin classes and applications to orbit spaces" (to appear in *Lecture Notes*, Springer Verlag) in which these connections between Topology and Number Theory are studied for higher dimensional manifolds and the appropriate generalizations of Dedekind sums. Part of this work was motivated by a formula of Bott on the rational Pontrjagin classes of complex projective space divided by an action of a finite group. A separate account of the number-theoretical part of Zagier's work will appear in *Mathematische Annalen.*

APPENDIX

Note 1 (see page 15). Set

$$\phi(\xi) = \left(\frac{\xi}{1-\xi} + \frac{1}{2}\right)\left(\frac{1}{\xi^h - 1} + \frac{1}{2}\right);$$

then (17) reads

$$s(h,k) = \frac{1}{k} \sum_{\xi}{}' \phi(\xi),$$

where the sum Σ'_{ξ} is extended over all kth roots of unity, except $\xi = 1$. If we multiply out the factors of $\phi(\xi)$, we obtain

$$\phi(\xi) = \frac{\xi}{(1-\xi)(\xi^h-1)} + \frac{\xi}{2(1-\xi)} + \frac{1}{2(\xi^h-1)} + \frac{1}{4},$$

so that

$$\frac{1}{k} \sum_{\xi}{}' \phi(\xi) = \frac{1}{k} \sum_{\xi}{}' \frac{\frac{1}{2}(1+\xi^{h+1})}{(1-\xi)(\xi^h-1)} + \frac{k-1}{4k}.$$

The last sum may also be written as

$$-\frac{1}{k}\sum_{\xi}{}' \frac{1+\frac{1}{2}(\xi^{h+1}-1)}{(\xi^h-1)(\xi-1)}$$

$$= -\frac{1}{k}\sum_{\xi}{}' \frac{1}{(\xi^h-1)(\xi-1)} - \frac{1}{2k}\sum_{\xi}{}' \frac{\xi^{h+1}-1}{(\xi^h-1)(\xi-1)}.$$

The desired assertion,

$$\text{(18a)} \qquad s(h,k) = -\frac{1}{k}\sum_{\xi}{}' \frac{1}{(\xi^h-1)(\xi-1)} + \frac{k-1}{4k}$$

is now seen to be equivalent to the equality

$$\sum_{\xi}{}' \frac{\xi^{h+1}-1}{(\xi^h-1)(\xi-1)} = 0.$$

To prove this, we write the summand successively as

$$\frac{1+\xi+\cdots+\xi^{h-1}+\xi^h}{\xi^h-1} = \frac{\xi^h}{\xi^h-1} + \frac{1+\xi+\cdots+\xi^{h-1}}{(\xi-1)(\xi^{h-1}+\cdots+\xi+1)}$$

$$= \frac{1}{1-\xi^{-h}} - \frac{1}{1-\xi} = \frac{1}{1-\xi_1} - \frac{1}{1-\xi},$$

with $\xi_1 = \xi^{-h}$. In the summation $\sum_{\xi}'$, ξ runs through all kth roots of unity except 1, and so does ξ_1. Consequently

$$\sum_{\xi}{}' \left(\frac{1}{1-\xi_1} - \frac{1}{1-\xi}\right) = 0$$

and the proof of (18a) is complete.

Note 2 (see page 22).

LEMMA 4. *Let $f(x)$, $g(x)$ and $q(x)$ be real valued functions of bounded variation in $a \leqq x \leqq b$, such that no two of them have any discontinuities in common. Then*

$$\int_a^b f(x)d(g(x)q(x)) = \int_a^b f(x)g(x)dq(x) + \int_a^b f(x)q(x)dg(x).$$

Proof. Let $f(x)$ and $\phi(x)$ be functions of bounded variation on $[a, b]$ and have no discontinuities in common; then

$$\int_a^b f(x)d\,\phi(x)$$

exists as a Stieltjes integral. The assumptions of the lemma insure the existence of each of the three integrals in (28).

Let $a = x_0 < x_1 < \cdots < x_j < \cdots < x_n = b$ be a partition of the interval $[a, b]$, select any points ξ_j so that $x_{j-1} \leqq \xi_j \leqq x_j$, and observe that

$$\begin{aligned} &\sum_{j=1}^{n} f(\xi_j)\{g(x_j)q(x_j) - g(x_{j-1})q(x_{j-1})\} \\ &= \sum_{j=1}^{n} f(\xi_j)g(x_j)\{q(x_j) - q(x_{j-1})\} \\ &\quad + \sum_{j=1}^{n} f(\xi_j)q(x_{j-1})\{g(x_j) - g(x_{j-1})\}. \end{aligned} \tag{72}$$

When we refine the partition, the limit approached by the first member is, by definition, the Stieltjes integral

$$\int_a^b f(x)d(g(x)q(x)).$$

Being of bounded variation, f, g, and h may each be written as the difference of two monotonically increasing functions, say

$f(x) = \Phi(x) - \phi(x)$, $g(x) = \Psi(x) - \psi(x)$, and

$$q(x) = X(x) - \chi(x).$$

If we substitute these expressions in the second member of (72), each of the two sums splits into eight similar sums, corresponding to the eight possible selections of capital and lower case ϕ, ψ, and χ in each product. Let us consider in some detail one of these sums, say

$$s_n = \sum_{j=1}^{n} \phi(\xi_j)\psi(x_j)\{\chi(x_j) - \chi(x_{j-1})\}.$$

By the monotonicity of the functions involved one has

$$\sum_{j=1}^{n} \phi(x_{j-1})\psi(x_j)\{\chi(x_j) - \chi(x_{j-1})\}$$

$$\leqq s_n \leqq \sum_{j=1}^{n} \phi(x_j)\psi(x_j)\{\chi(x_j) - \chi(x_{j-1})\}.$$

When the partition is refined, the first and last member converge both to

$$\int_a^b \phi(x)\psi(x)d\chi(x) \tag{73}$$

and so does s_n. The other sums are handled in the same way. It now follows, first, that the second member of (72) approaches a limit. Next, we see that the limit of each of the two last sums in (72) is represented by eight Stieltjes integrals of which (73) is typical. The first four integrals that we obtain from the first sum in the second member of (72) are

$$\int_a^b \Phi(x)\Psi(x)dX(x) - \int_a^b \Phi(x)\Psi(x)d\chi(x)$$

$$- \int_a^b \Phi(x)\psi(x)dX(x) + \int_a^b \Phi(x)\psi(x)d\chi(x).$$

These integrals can be recombined and we obtain

$$\int_a^b \Phi(x)\Psi(x)d(X(x)-\chi(x)) - \int_a^b \Phi(x)\psi(x)d(X(x)-\chi(x))$$

$$= \int_a^b \Phi(x)(\Psi(x)-\psi(x))d(X(x)-\chi(x)) = \int_a^b \Phi(x)g(x)dq(x).$$

Similarly, the other four integrals combine into $-\int_a^b \phi(x)g(x)dq(x)$. It follows that the limit of the first sum in the second member of (72) is

$$\int_a^b (\Phi(x)-\phi(x))g(x)dq(x) = \int_a^b f(x)g(x)dq(x).$$

The eight integrals, whose sum equals the limit (under arbitrary refinement of the partition) of the last sum in (72) are of essentially the same form as the preceding ones, except that the Ψ's and χ's are interchanged. We handle them in the same way, the result is

$$\int_a^b f(x)q(x)dg(x),$$

and this finishes the proof of Lemma 4.

Note 3 (see page 41).

$$A+4 = \sum_{x,y,z} \left(\frac{x}{a}+\frac{y}{b}+\frac{z}{c}-2\right)^2$$

$$= \sum_{x,y,z} \left\{\frac{x^2}{a^2}+\frac{y^2}{b^2}+\frac{z^2}{c^2}+2\left(\frac{xy}{ab}+\frac{yz}{bc}+\frac{zx}{ca}\right) - 4\left(\frac{x}{a}+\frac{y}{b}+\frac{z}{c}\right)+4\right\}.$$

Next

$$\sum_{0 \leq x \leq a-1} \frac{x^2}{a^2} = \frac{1}{a^2} \frac{(a-1)a(2a-1)}{6} = \frac{(a-1)(2a-1)}{6a},$$

so that

$$\sum_{x,y,z} \frac{x^2}{a^2} = \frac{(a-1)(2a-1)}{6a} bc.$$

The contributions of $\Sigma_{x,y,z}\, y^2/b^2$ and $\Sigma_{x,y,z}\, z^2/c^2$ are obtained by circular permutations of a, b, and c, and adding the results one obtains

$$abc - \frac{1}{2}(ab + bc + ca) + \frac{1}{6}\left(\frac{bc}{a} + \frac{ca}{b} + \frac{ab}{c}\right).$$

Similarly,

$$\sum_{x} \frac{x}{a} = \frac{1}{a}\frac{(a-1)a}{2} = \frac{a-1}{2}, \quad \sum_{x,y,z} \frac{x}{a} = \frac{1}{2} bc(a-1),$$

so that

$$-4 \sum_{x,y,z} \left(\frac{x}{a} + \frac{y}{b} + \frac{z}{c}\right) = -2\{3abc - (bc + ca + ab)\},$$

and also $4\, \Sigma_{x,y,z} 1 = 4abc$.

Finally, $\Sigma_z\, (xy/ab) = (c/ab)xy$, so that

$$\sum_{x,y,z} \frac{xy}{ab} = \frac{c}{ab}\frac{a(a-1)}{2}\frac{b(b-1)}{2} = \frac{1}{4}c(a-1)(b-1),$$

and

$$2 \sum_{x,y,z} \left(\frac{xy}{ab} + \frac{yz}{bc} + \frac{zx}{ca}\right) = \frac{1}{2}\{3abc + a + b + c - 2(ab + bc + ca)\}.$$

By adding these results we obtain

$$A + 4 = abc - \frac{1}{2}(ab + bc + ca) + \frac{1}{6}\left(\frac{bc}{a} + \frac{ca}{b} + \frac{ab}{c}\right)$$
$$+ \frac{1}{2}(a + b + c) - 6abc + 2(ab + bc + ca)$$
$$+ 4abc + \frac{3}{2}abc - (ab + bc + ca)$$
$$= \frac{1}{2}abc + \frac{1}{2}(ab + bc + ca) + \frac{1}{6}\left(\frac{bc}{a} + \frac{ca}{b} + \frac{ab}{c}\right)$$
$$+ \frac{1}{2}(a + b + c),$$

as claimed.

Note 4 (see page 43).

THEOREM A. (see [**46**]). *If* $(a,b) = (b,c) = (c,a) = 1$, *then*

$$(74)\quad \left(s(bc,a) - \frac{bc}{12a}\right) + \left(s(ca,b) - \frac{ca}{12b}\right) + \left(s(ab,c) - \frac{ab}{12c}\right)$$
$$\equiv -\frac{1}{4} - \frac{abc}{12} + \frac{1}{12abc} \pmod 2.$$

This theorem follows easily from the Reciprocity Theorem (4) for Dedekind sums and the following:

THEOREM B. (see [**46**]). *If* $(a,b) = (b,c) = (c,a) = 1$, *then*

$$(75)\quad \left(s(bc,a) - \frac{bc}{12a}\right) + \left(s(ca,b) - \frac{ca}{12b}\right) - \left(s(c,ab) - \frac{c}{12ab}\right)$$
$$+ \frac{abc}{12} \equiv 0 \pmod 2.$$

Proof of Theorem A. By (4),

$$s(c,ab) + s(ab,c) = -\frac{1}{4} + \frac{1}{12}\left(\frac{c}{ab} + \frac{1}{abc} + \frac{ab}{c}\right),$$

so that, if we replace $s(c,ab)$, (75) becomes precisely (74) as claimed.

It remains to prove Theorem B.

Proof of Theorem B. The proof of this theorem is rather lengthy. In order to cut down on case distinctions we set again $\theta = (3,k)$, so that $\theta = 3$ if $3 \mid k$, $\theta = 1$ otherwise. It follows that $\theta k = 3k$ and is divisible by 9 if $3 \mid k$; otherwise, $\theta k = k$.

LEMMA 9. (75) *holds if a and b are odd.*

We shall assume for a moment Lemma 9 and use it to prove Theorems A and B together. The main difficulty comes from the fact that, while (74) is symmetric in a, b, and c, (75) is not. The purpose of the following, somewhat subtle, reasoning is to overcome that difficulty. The proof will be completed by an independent proof of Lemma 9. Clearly, Lemma 9 is just Theorem B with the added assumption that a and b are odd. Since a, b, and c are coprime in pairs, at least two of them are odd. Let us assume for a moment that these are a and b; then Lemma 9 shows that Theorem B holds, and we just saw that this implies Theorem A. Formula (74), however, is symmetric in all three letters a, b, and c and holds, therefore, provided that they are coprime in pairs and any two of them are odd. The last condition, however, is already implied by the first (i.e., by the pairwise coprimality), so that $(a,b) = (b,c) = (c,a) = 1$ is sufficient to insure the validity of Theorem A. Replacing in (74) any of the occurring Dedekind sums with the help of the reciprocity law, we obtain (75), either as written, or with the letters a, b, c permutated cyclically. It follows that (75) and the relations that we obtain from it by permuting a, b and c also hold, so that although (75) is not symmetric in

a, b, c, Theorem B holds for all triplets of integers that are coprime in pairs. It is now clear that by proving Lemma 9, the proof of both Theorems A and B will be complete.

Proof of Lemma 9. The statement is equivalent to

$$12ab\,s(bc,a) + 12ab\,s(ca,b) - 12ab\,s(c,ab) - b^2c - ca^2 + c + a^2b^2c \equiv 0 \pmod{24ab}.$$

We denote the sum of the first three terms by D and observe that the others are $c(a^2b^2 - a^2 - b^2 + 1) = c(a^2-1)(b^2-1)$, so that we have to prove that

$$D + c(a^2-1)(b^2-1) \equiv 0 \pmod{24\,ab}. \tag{76}$$

We recall that $3ab$ is odd, so that $(3ab, 8) = 1$, and it is sufficient to show that

$$D \equiv -c(a^2-1)(b^2-1) \pmod{3ab} \tag{77}$$

and

$$D \equiv -c(a^2-1)(b^2-1) \pmod{8} \tag{78}$$

both hold.

The congruence (78) can be disposed of immediately. Indeed, $(a^2-1)(b^2-1) \equiv 0 \pmod 8$, because a and b are odd, so that (78) reduces to $D \equiv 0 \pmod 8$. Also, $12a\,s(bc,a)$ and the other two summands of D that contain Dedekind sums can be replaced modulo 8 by (42), and so we obtain

$$\begin{aligned} D &\equiv b\left(a + 1 - 2\left(\frac{bc}{a}\right)\right) + a\left(b + 1 - 2\left(\frac{ca}{b}\right)\right) - \left(ab + 1 - 2\left(\frac{c}{ab}\right)\right) \\ &\equiv ab + b + a - 1 + 2\left\{\left(\frac{c}{ab}\right) - b\left(\frac{bc}{a}\right) - a\left(\frac{ca}{b}\right)\right\} \\ &\equiv 2ab - (a-1)(b-1) + Q \pmod 8. \end{aligned} \tag{79}$$

Here $Q = 2\{(c/a)(c/b) - b(b/a)(c/a) - a(c/b)(a/b)\} = 2\{(c/a) - a(a/b)\}\{(c/b) - b(b/a)\} - 2ab(a/b)(b/a)$. The expressions in braces on the right hand side are both even, because a and b and the Jacobi symbols are all odd. Consequently, the first summand on the right hand side vanishes (mod 8) and (79) becomes

$$D \equiv 2ab\left\{1 - \left(\frac{a}{b}\right)\left(\frac{b}{a}\right)\right\} - (a-1)(b-1)$$

$$\equiv 2ab\,R - (a-1)(b-1) \pmod 8,$$

say. By the quadratic reciprocity law,

$$\left(\frac{a}{b}\right)\left(\frac{b}{a}\right) = (-1)^{(a-1)(b-1)/4},$$

and R vanishes if either $a \equiv 1 \pmod 4$ or $b \equiv 1 \pmod 4$; otherwise, if $a \equiv b \equiv 3 \pmod 4$, $R = 2$. It follows that always

$$\tfrac{1}{2}\{1 - (-1)^{(a-1)(b-1)/4}\} \equiv \tfrac{1}{4}(a-1)(b-1) \pmod 2,$$

and (79) becomes $D \equiv 4ab \cdot \frac{1}{4}(a-1)(b-1) - (a-1)(b-1) \equiv (ab-1)(a-1)(b-1) \equiv 0 \pmod 8$, because all three factors are even. This finishes the proof of (78).

It still remains to prove (77). The essential element of this proof is contained in

LEMMA 10. *Let $(h,k) = 1$; then*

$$12hk\,s(h,k) \equiv (1-k^2)(1+h^2) \pmod{3k} \tag{80}$$

and

$$12hk\,s(k,h) \equiv k^2(h^2-1) \pmod{3k}. \tag{81}$$

Proof of Lemma 10. We shall make use of the following congruences:

$$(82) \qquad 12hk\,s(h,k) \equiv h^2 + 1 \pmod{\theta k};$$

$$(83) \qquad 12\,k\,s(h,k) \equiv 0 \pmod 3 \text{ if and only if } 3 \nmid k;$$

$$(84) \qquad 12hk\,s(k,h) \equiv 0 \pmod{\theta k};$$

$$(85) \qquad 12k\,s(h,k) \equiv h(k^2 - 1) \pmod 3.$$

Formula (82) is the same as (36) and has been proved.

Next, (39) shows that $12k\,s(h,k) \equiv 2h(k-1)(2k-1)$ (mod 3) and (83) immediately follows.

We also observe that, on account of Theorem 2, the denominator of $s(k,h)$ is a divisor of $2h\theta'$, where $\theta' = (3,h)$. In particular, $12\,h\,s(k,h)$ is always an integer, so that

$$(86) \qquad 12\,hk\,s(k,h) \equiv 0 \pmod k.$$

If $3 \nmid k$, then $\theta = 1$, $\theta k = k$ and (84) holds, because it reduces to (86). If $3 \mid k$, then $\theta k = 3k$, but then $3 \nmid h$, $\theta' = 1$, and $2h\,s(k,h) = m$ is an integer, by Theorem 2. Consequently,

$$12\,hk\,s(k,h) = 6km$$

and

$$6km \equiv 0 \pmod{3k}$$

so that (84) again holds.

Finally (85) follows directly from the definition of the Dedekind sums:

$$12k\,s(h,k) = 12\sum_{\mu=1}^{k}\mu\left(\left(\frac{h\mu}{k}\right)\right) = 12\sum_{\mu=1}^{k-1}\mu\left(\frac{h\mu}{k} - \left[\frac{h\mu}{k}\right] - \frac{1}{2}\right)$$

$$= 2h(k-1)(2k-1) - 12\sum_{\mu=1}^{k-1}\mu\left[\frac{h\mu}{k}\right] - 3k(k-1)$$

$$\equiv 2h(k-1)(2k-1) \pmod 3$$

$$\equiv -h(k-1)(-k-1) \equiv h(k^2-1) \pmod 3.$$

Lemma 10 now follows immediately. Indeed, let $3\,|\,k$, so that $3\nmid h$ and $\theta = 3$. Then $1 - k^2 \equiv 1 \pmod{3k}$, and (80) holds by (82). Also, $h^2 \equiv 1 \pmod 3$, so that $k^2(h^2-1) \equiv 0 \pmod{3k}$, and (81) holds by (84). If $3\nmid k$, $\theta = 1$ and both, (80) and (81) have to be shown to hold separately (mod k) and (mod 3). (80) holds (mod k) by (82) and (mod 3) by (83), on account of $k^2 - 1 \equiv 0 \pmod 3$. (81) holds trivially (mod k) because both sides actually vanish (mod k); and it holds (mod 3) by (85), after h and k have been interchanged. The Lemma is proved.

We now proceed to prove (77). Once this is accomplished, (76) follows on account of (78) and this finishes the proof of Lemma 9, hence that of Theorems A and B.

We have, successively,

$$A = 12abc\,s(bc,a) \equiv (1-a^2)(1+b^2c^2) \pmod{3a} \text{ by (80)},$$

$$B = 12abc\,s(ca,b) \equiv c^2a^2(b^2-1) \pmod{3ac} \text{ by (81)},$$

$$C = 12abc\,s(c,ab) \equiv (1-a^2b^2)(1+c^2) \pmod{3ab} \text{ by (80)}.$$

In particular, all congruences hold mod $3a$, so that

$$A + B - C = cD \equiv a^2b^2 + b^2c^2 - c^2a^2 - a^2 - c^2 + a^2b^2c^2 \pmod{3a}.$$

We also observe that

$$F = 2a^2b^2c^2 - a^2 - 2a^2c^2 + a^2b^2 = a^2(b^2-1)(2c^2+1)$$

vanishes modulo $3a$.

Indeed, $3 \mid b^2-1$ unless $3 \mid b$. In that case, however, $3 \nmid c$ and $2c^2+1 \equiv 0 \pmod 3$. It follows that

$$cD \equiv cD - F \equiv -c^2(a^2-1)(b^2-1) \pmod{3a}.$$

Here both members of the congruence are symmetric in a and b, so that one also has $cD \equiv -c^2(a^2-1)(b^2-1) \pmod{3b}$. From $(a,b) = 1$, it now follows that

$$cD \equiv -c^2(a^2-1)(b^2-1) \pmod{3ab}.$$

If $3 \nmid c$, a factor c may be cancelled, and (77) is proved. If, however, $3 \mid c$, then one can only infer that

$$D \equiv -c(a^2-1)(b^2-1) \pmod{ab}.$$

Now, however, $3 \nmid ab$, and in order to complete the proof of (77) it only remains to verify that (77) holds also mod 3. On account of $3 \mid c$, we have to verify only that $D \equiv 0 \pmod 3$. This, however, is indeed the case, because the conditions $3 \nmid a$ and $3 \nmid b$ imply by Theorem 2 that $2a\,s(bc,a)$, $2b\,s(ca,b)$, and $2ab\,s(c,ab)$ are all integers, so that all three summands of D are divisible by 3. This finishes the proof of (77), hence that of Lemma 9 and of Theorems A and B.

REFERENCES

1. T. M. Apostol, Generalized Dedekind sums and transformation formulae of certain Lambert series, Duke Math. J., 17 (1950) 147–157.

2. ———, Theorems on generalized Dedekind sums, Pacific J. Math., 2 (1952) 1–9.

3. P. Bachmann, Die Elemente der Zahlentheorie, Teubner, Leipzig, 1892.

4. K. Barner, Über die Werte der Ringklassen-L-Funktionen reell-quadratischer Zahlkörper an natürlichen Argumentstellen, J. Number Theory, 1 (1969) 28–64.

5. L. Carlitz, Some theorems on generalized Dedekind sums, Pacific J. Math., 3 (1953) 513–522.

6. ———, The reciprocity theorem for Dedekind sums, Pacific J. Math., 3 (1953) 523–527.

7. ———, Dedekind sums and Lambert series, Proc. Amer. Math. Soc., 5 (1954) 580–584.

8. ———, A note on generalized Dedekind sums, Duke Math. J., 21 (1954) 399–404.

9. ———, Some finite summation formulas of arithmetic character, Publ. Math., Debrecen, 6 (1959) 262–268.

10. ———, Generalized Dedekind sums, Math. Z., 85 (1964) 83–90.

11. ———, A theorem on generalized Dedekind sums, Acta Arith., 11 (1965) 253–260.

12. ———, Linear relations among generalized Dedekind sums, J. Reine Angew. Math., 280 (1965) 154–162.

13. ———, A three-term relation for Dedekind-Rademacher sums, Publ. Math., Debrecen, 14 (1967) 119–124.

14. R. Dedekind, Erläuterungen zu zwei Fragmenten von Riemann — Riemann's Gesammelte Math. Werke, 2nd edition, 1892, pp. 466–472; also Dedekind's Gesammelte Math. Werke, 1930, vol. 1, pp. 159–173.

15. ———, Schreiben an Herrn Borchardt, J. Reine Angew. Math., 83 (1877) 265–292; also Dedekind's Gesammelte Math. Werke, vol. 1, pp. 174–201.

16. U. Dieter, Zur Theorie der Dedekindschen Summen, Inauguraldissertation, Kiel, 1957, mimeographed (especially p. 15 and p. 52).

17. U. Dieter, Beziehungen zwischen Dedekindschen Summen, Abh. Math. Sem. Univ. Hamburg, 21 (1957) 109–125.

18. ———, Das Verhalten der Kleinschen Funktionen log $\sigma_{g,h}(\omega_1, \omega_2)$ gegenüber Modultransformationen und verallgemeinerte Dedekindsche Summen, J. Reine Angew. Math., 201 (1959) 37–70.

19. ———, Autokorrelation multiplikativ erzeugter Pseudo-Zufallszahlen — I. Oberwolfach-Tagung über Operations Research — Verlag Anton Hain, Meisenheim, pp. 69–85.

20. U. Dieter and J. Ahrens, An exact determination of serial correlations of pseudo-random numbers, Numer. Math., 17 (1971) 101–123.

21. U. Dieter, J. Ahrens and A. Grube, Pseudo-random numbers: A new proposal for the choice of multiplicators, Computing 6 (1970) 121–138.

22. E. Grosswald, Topics from the theory of numbers, Macmillan, New York, 1966.

23. G. H. Hardy and E. M. Wright, An Introduction to the Theory of Numbers, 3rd edition, Clarendon Press, Oxford, 1954.

24. E. Hecke, Über die Kroneckersche Grenzformel für reelle quadratische Körper und die Klassenzahl relativ-Abelscher Körper, Ges. Werke, pp. 198–207.

25. S. Iseki, The transformation formula for the Dedekind modular function and related functional equations, Duke Math. J., 24 (1957) 653–662.

26. B. Jansson, Random Number Generators, Almqvist and Wiksell, Stockholm, 1966.

27. D. E. Knuth, The Art of Computer Programming, Addison-Wesley, Reading, Mass., 1968.

28. E. Landau, Vorlesungen über Zahlentheorie, Chelsea, New York, vol. 2; also Nachr. Akad. Wiss. Göttingen Math.-Phys. Kl., 1924, p. 203.

29. J. Lehner, A partition function connected with the modulus five, Duke Math. J., 8 (1941) 631–655.

30. J. Livingood, A partition function with the prime modulus $p > 3$, Amer. J. Math., 67 (1945) 194–208.

31. H. Maass, Lectures on modular functions of one complex variable, Tata Institute of Fundamental Research, Bombay, 1964.

32. C. Meyer, Über einige Anwendungen Dedekindscher Summen, J. Reine Angew. Math., 198 (1957) 143–203.

32a. C. Meyer, Bemerkungen zu den allgemeinen Dedekindschen Summen, J. Reine Angew. Math., 205 (1960) 186–196.

33. ———, Die Berechnung der Klassenzahl Abelscher Körper über quadr. Zahlkörpern, Akademie-Verlag, Berlin, 1957.

34. M. Mikolás, On certain sums generating the Dedekind sums and their reciprocity laws, Pacific J. Math., 7 (1957) 1167–1178.

35. ———, Über gewisse Lambertsche Reihen, I: Verallgemeinerung der Modulfunktion $\eta(\zeta)$ und ihrer Dedekindschen Transformationsformel, Math. Z., 68 (1957) 100–110.

36. L. J. Mordell, On the reciprocity formula for Dedekind sums, Amer. J. Math., 73 (1951) 593–598.

37. ———, Lattice points in a tetrahedron and generalized Dedekind sums, J. Indian Math. Soc., 15 (1951) 41–46.

38. M. Newman, Private, oral communication to Professor H. Rademacher.

39. H. Rademacher, Zur Theorie der Modulfunktionen, Atti del Congr. Intern. dei Mat., Bologna, (1928) 297–301.

40. ———, Zur Theorie der Modulfunktionen, J. Reine Angew. Math., 167 (1931) 312–366.

41. ———, Eine arithmetische Summenformel, Monatshefte für Math. u. Phys., 39 (1932) 221–228.

42. ———, Bestimmung einer gewissen Einheitswurzel in der Theorie der Modulfunktionen, J. London Math. Soc., 7 (1931) 14–19.

43. ———, Egy Reciprocitásképletröl a Modulfüggevények Elméletéböl, Mat. Fiz. Lapok, 40 (1933) 24–34.

44. H. Rademacher and A. Whiteman, Theorems on Dedekind sums, Amer. J. Math., 63 (1941) 377–407.

45. H. Rademacher, Die Reziprozitätsformel für Dedekindsche Summen, Acta Sci. Math. (Szeged), 12 (B) (1950) 57–60.

46. ———, On Dedekind sums and lattice points in a tetrahedron, Studies in mathem. and mechanics presented to R. von Mises (1954), pp. 49–53.

47. ———, Generalization of the reciprocity formula for Dedekind sums, Duke Math. J., 21 (1954) 391–398.

48. ———, On the transformation of log $\eta(\zeta)$, J. Indian Math. Soc., 19 (1955) 25–30.

49. H. Rademacher, Zur Theorie der Dedekindschen Summen, Math. Z., 63 (1956) 445–463.

50. ———, Some remarks on certain generalized Dedekind sums, Acta Arith., 9 (1964) 97–105.

51. ———, A convergent series for the partition function $p(n)$, Proc. Nat. Acad. Sci. USA, 23 (1937) 78–84.

52. ———, On the partition function $p(n)$, Proc. London Math. Soc., (2), 43 (1937) 241–254.

53. ———, On the expansion of the partition function in a series, Ann. Math., (2), 44 (1943) 416–422.

54. L. Rédei, Bemerkung zur vorstehenden Arbeit des Herrn H. Rademacher, Mat. Fiz. Lapok, 40 (1933) 35–39.

55. ———, Elementarer Beweis und Verallgemeinerung einer Reziprozitätsformel von Dedekind, Acta Sci. Math. (Szeged), 12 (B) (1950) 236–239.

56. G. J. Rieger, Dedekindsche Summen in algebr. Zahlkörpern, Math. Ann., 141 (1960) 377–383.

57. B. Riemann, Fragmente über die Grenzfälle der elliptischen Modulfunktionen — Gesammelte Math. Werke, Dover, New York, 1953, pp. 445–465.

58. M. Riesz, Sur le lemme de Zolotareff et sur la loi de réciprocité des restes quadratiques, Math. Scand., 1 (1955) 159–169.

59. H. Salié, Zum Wertevorrat der Dedekindschen Summen, Math. Z., 72 (1959) 61–75.

60. E. Schering, Zur Theorie der quadratischen Reste, Acta Math., 1 (1882) 153–170.

61. C. L. Siegel, A simple proof of $\eta(-1/\tau)=\eta(\tau)\sqrt{\tau/i}$, Mathematika, 1 (1954) 4.

62. B. Schoeneberg, Verhalten von speziellen Integralen 3. Gattung bei Modultransformationen und verallgemeinerte Dedekindsche Summen — Abh. Math. Sem. Univ. Hamburg, 30 (1967) 1–10.

63. K. Wohlfart, Über Dedekindsche Summen und Untergruppen der Modulgruppe, Abh. Math. Sem. Univ. Hamburg, 23 (1959) 5–10.

64. E. Zolotareff, Nouvelle démonstration de la loi de réciprocité de Legendre, Nouvelles Annales de Math., (2), 11 (1872) 355–362.

LIST OF THEOREMS AND LEMMAS

NAME INDEX

SUBJECT INDEX

For four subjects that occur very frequently (e. g., Dedekind sums), only the page of first occurrence and / or of the definition is given followed by the indication *et seq.*